George Francis Atkinson

The study of the biology of ferns by the collodion method

for advanced and collegiate students

George Francis Atkinson

The study of the biology of ferns by the collodion method
for advanced and collegiate students

ISBN/EAN: 9783742800428

Manufactured in Europe, USA, Canada, Australia, Japa

Cover: Foto ©Klaus-Uwe Gerhardt /pixelio.de

Manufactured and distributed by brebook publishing software
(www.brebook.com)

George Francis Atkinson

The study of the biology of ferns by the collodion method

THE BIOLOGY OF FERNS BY THE
COLLODION METHOD

The Study of

The Biology of Ferns

By the

Collodion Method

FOR ADVANCED AND COLLEGIATE STUDENTS

BY

GEORGE F. ATKINSON, Ph.B.

ASSOCIATE PROFESSOR OF CRYPTOGAMIC BOTANY IN CORNELL UNIVERSITY

New York

MACMILLAN AND CO.

AND LONDON

1894

49606

PREFACE.

THE author was led to the preparation of this little work by the success which attended his efforts in applying in his classes the "collodion method" to the preparation of the very delicate tissues of ferns, and especially to the infiltration of prothallia, without shrinkage. A class of nine students sectioned prothallia with ease and accuracy, making permanent microscopic preparations showing various stages of development of the sexual organs and the embryo, besides good series showing the development of the sporangia and spores. The first suggestion was the preparation of a simple laboratory guide giving directions for preparing the various tissues, accompanying this with a few illustrations, so far as possible made from preparations put up by the students in their regular work, and some descriptive matter which would be helpful to the student in mastering the detail of structure and development.

It seemed desirable to the author that some features in the development not represented by specimens selected from the work of the students should be included. Accordingly, he was led to a protracted study of various phases of development, since it was desirable that all illustrations should be original, and that those of sections should be from preparations put up by this method. The work thus grew, since several questions arose which it seemed important to enquire into quite closely. Especially the question of the structure of the annulus in the various orders of ferns, and the function of the different parts of the sporangium in the dispersion of the spores, called for study.

The author does not wish to insist on the merits of the collodion method in comparison with the paraffin method of infiltrating plant tissues, although perhaps the very delicate tissues will suffer less from treatment by the former method. Those who have been accustomed to the paraffin method can doubtless use it to good purpose in connection with the practical part. Therefore the methods described in Part II. for the preparation of the tissues for infiltration are not inseparable from the collodion method, but will, perhaps, serve as well to prepare the tissues for paraffin infiltration.

The fact that all the illustrations are original and drawn from preparations put up according to a method in plant histology within the reach of all students of botany, it is hoped, will give a stimulus to many to enquire into the mysteries of fern development, and the encouragement that like results will certainly attend them.

Where the illustrations are made from preparations put up by the students, credit is given to the preparator. All the other illustrations are from preparations made by the author.

Ferns, from their intermediate position in the plant kingdom, as well as from their varied beauty of graceful form, are objects of great interest to the student of nature. The differential characters found in their well-developed tissues and in the specialization of their leaves and fruit structures, offer a wide field for observation and discrimination to those interested in questions of taxonomy. But the subject appeals especially to the student of biology, since the great phylogenetic problem of the line of succession reaching from the lower Thallophyta, as they are sometimes called, to the Spermaphyta, is being traced through this group and its near relatives.

The study of the biology of ferns is, therefore, of double importance. Not only does it offer to the student of plant biology a field in which the phenomena of plant development and plant structure come in touch, on the one hand, with groups of extremely simple manifestations of life and form, and, on the other hand, with those of great complexity ; but it also offers inducements in the way of return for research. For great as is the work which has been accomplished in studies of development and in tracings of phylogeny, the life histories of but few individuals have been accurately traced, while the line of descent needs to be more clearly unveiled by farther careful studies of life histories.

The present work relates chiefly to the homosporous Leptosporangiate Filicineæ, and seven chapters of descriptive matter are devoted to them. One chapter is added on the Ophioglosseæ, not so much because it is held that they stand next the former in the natural series, but because it is quite easy to obtain fresh material for study; its members present excellent subjects for comparative study, and they are popularly known as ferns.

CORNELL UNIVERSITY, ITHACA, N.Y.,
December, 1893.

CONTENTS.

———◦◦◦———

PART I. DESCRIPTIVE.

X CONTENTS.

Part I.

DESCRIPTIVE.

THE BIOLOGY OF FERNS.

GAMETOPHYTE.

Life Cycle of Ferns. — The normal life cycle of ferns consists of two successive or complemental phases: the gametophyte (prothalline stage, oophyte) and sporophyte (fern plant). The gametophyte is the earliest form to appear in the life cycle, and probably also was the first in time to appear on the earth. It is usually a flat expanded plate of tissue, or it may be filamentous and alga-like. It is termed the prothallium. It proceeds from the germinating spore, and bears the sexual organs of reproduction. As a result of fertilization the egg cell develops into the fern plant with which we are all so familiar, the sporophyte. This in turn bears the spores, and is usually perennial. The elimination of one or the other of these phases from the normal life cycle, or the production of new gametophytes or sporophytes as buds, frequently happens, and will be referred to in detail hereafter.

With this preliminary general statement we will proceed to discuss more in detail the various phenomena of development and the special conformation of the different phases.

Since the prothallium is a result of growth from the germinating spore, we may look at the characters of the spores in general, while their development will be reserved for study in connection with the sporophyte.

Spores. — The spores of ferns in form are of two different types, resulting from two ways of development. In most of

3

the Polypodiaceæ, Gleicheniaceæ, and Schizæaceæ the form approaches that of a quadrant of a sphere, and they are said to be bilateral or radial.   The spore usually consists of two

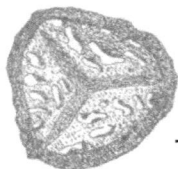

coats, an inner coat, the endospore, which surrounds the protoplasm, and is quite thin compared with the outer coat, the exospore.   The exospore is usually cuticularized and coloured, being also variously marked with warts, papillae, or ridges. In a pretty green-house fern, *Niphobolus crispum corymbiferum*, a native of Japan, the spores are bright yellow, and marked with stout, irregular warts.   In various species of *Aspidium*, the shield ferns, the

Fig. 1.   Spore of Pteris serrulata, showing the roughened exospore and the three-rayed fissure. Magnified 30 times more than the scale; scale = 1 mm.

exospore is produced into prominent wings which anastomose irregularly.   In *Polypodium vulgare* the elevations of the surface of the exosporium are in the form of short stout papillæ,

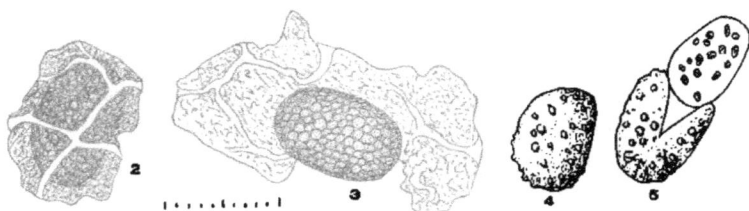

Fig. 2.   Spore of Aspidium acrostichoides with winged exospore. — Fig. 3.   Same crushed to remove exospore, and showing the endospore. — Figs. 4 and 5.   Spores of Asplenium filix-fœmina, showing same conditions.   All magnified 30 times more than the scale; scale = 1 mm.

while in *Cystopteris bulbifera* they are nearly cylindrical and quite long.   In *Schizæa pusilla* the surface of the exospore is beautifully marked, reminding one of the tessellated marking on the frustum of some diatom.

In some cases in the above-named families, and generally in the Cyatheaceæ, Hymenophyllaceæ, and Osmundaceæ, the spores are rounded or tetrahedral in shape.

Germination of Spores. — With the exception of the Hymeno-phyllaceæ and Osmundaceæ, the spores are long-lived, and require a period of rest before they will germinate, several species having been known to germinate after storage in the herbarium for several years. Having passed this period, usually a few months, they will germinate in a few days' time if placed under favourable conditions. If placed on moist soil, or some porous substratum which will retain a moist surface when protected from above, the coats of the spore absorb water so that it comes in contact with the protoplasm within. Some substance in the protoplasm possesses a strong avidity for water, which is then drawn in with great force. The protoplasmic lining of the cell will not permit water to filter out, and thus the strong endosmotic pressure exerted causes the bulk of the protoplasm to increase. This produces a powerful tension upon the two coats of the spore. The exospore cannot stretch to accommodate the increasing bulk of the protoplasm, and therefore it bursts. In a number of cases the endospore bursts also, and an entirely new wall of cellulose is deposited about the protoplasm.

Campbell found this to be the case in *Onoclea*. According to Rauwenhoff it occurs in the Gleicheniaceæ. If the stout exospore did not cling so tenaciously in many other ferns, this might be observed in quite a large number of cases probably. In most cases in the Polypodiaceæ after the exo-spore has burst, the cell elongates, producing a short pro-tonemal thread containing chlorophyll and starch grains in the protoplasm. It thus issues from the fissure in the exospore, and soon divides by a cross wall into two cells, the proximal cell containing less chlorophyll and giving rise to the first rhizoid. The rhizoid is an elongated, unsegmented, slender cell, usually destitute of chlorophyll. It may be entirely colourless, or in some cases assumes a brown colour. It en-ters the soil or other nutritive substratum on which the spore is, and functions as an absorbing organ to supply the young prothallium with food. It describes a sinuous course, and the free end is often somewhat flattened and irregular in outline, which thus forms a suitable hold for adherent particles, from

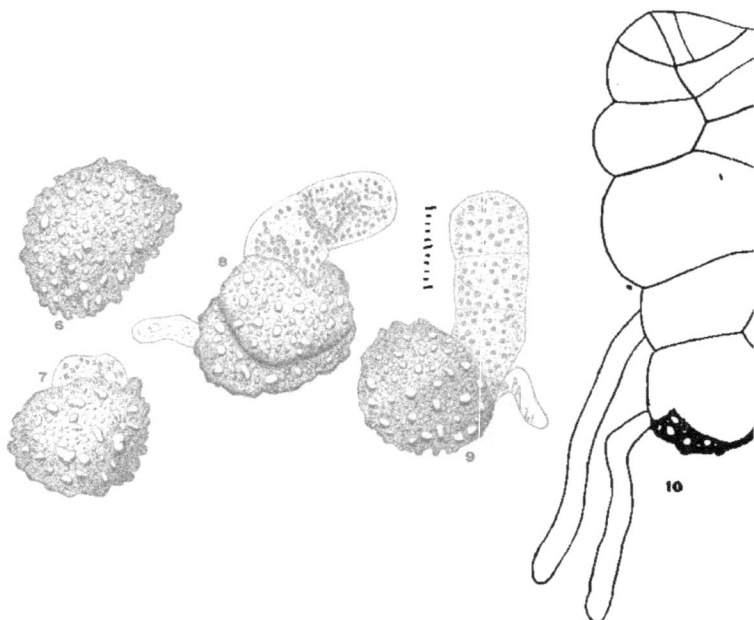

Fig. 6. Spore of Niphobolus crispum corymbiferum. — Fig. 7. Same in early stage of germination, showing first cell of protonemal thread. — Figs. 8 and 9. Farther development, showing young rhizoid and protonemal thread of two and three cells. — Fig. 10. Still older prothallium, showing the beginning of the expanded plate of cells, with the wedge-shaped apical cell at the growing end. A portion of the exospore still clings to the first cell of the prothallium. All the figures magnified 30 times more than the scale; scale = 1 mm.

which moisture and nutritive substances are drawn up much as in the root hairs of higher plants.

Prothallium. — The terminal cell of this young protonemal thread continues to elongate and divide by transverse walls, until a thread of several cells' length is developed, when under normal conditions growth takes place laterally as well as in a direct

line.   An expanded plate of cells is thus formed.   This is usually introduced by an oblique wall across the terminal cell of

the protonemal thread, followed by another oblique
wall forming an angle with
the first.   This starts a
wedge-shaped apical cell.
For some time successive
oblique walls are formed
across this growing apical
cell, and transverse walls
in the lateral cells produce
the plate of a single layer
of cells.   The beginning of
the plate of cells is not,
however, always introduced
by oblique division resulting in an apical cell, but
sometimes by a perpen-
dicular wall, thus forming
quadrangular apical cells.
For sometime this apical
cell is in advance of the
lateral growth, but soon

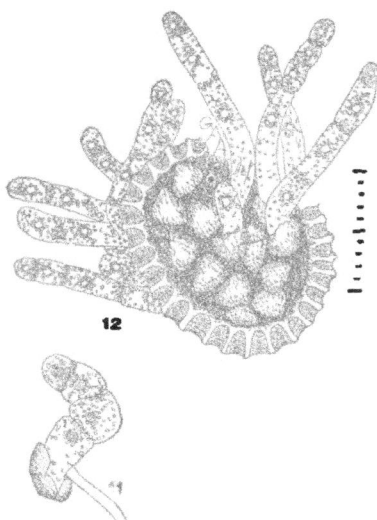

Figs. 11 and 12. Germinating spores of Pteris
serrulata.   In Fig. 12 the spores were still
within the sporangium, the sporangium having been sown at the time of the sowing of
the spores. Magnified 6 times more than
the scale; scale = 1 mm.

the tissue on either side extends in advance of the apical cell,
so that the young prothallium is heart-shaped.

Before long, growth takes place in a direction perpendicular
to the surface of the prothallium at the sinus, so that a cushion
of tissue several cells in depth is formed at this point.   Thus a
merismatic tissue, composed of small cells richly filled with
protoplasm, takes the place of the apical cell.   Meanwhile other
rhizoids are developed from the under side of the prothallium
at the posterior end, which give it a firm hold upon the
substratum.

In the development of the meristematic cushion, in place of

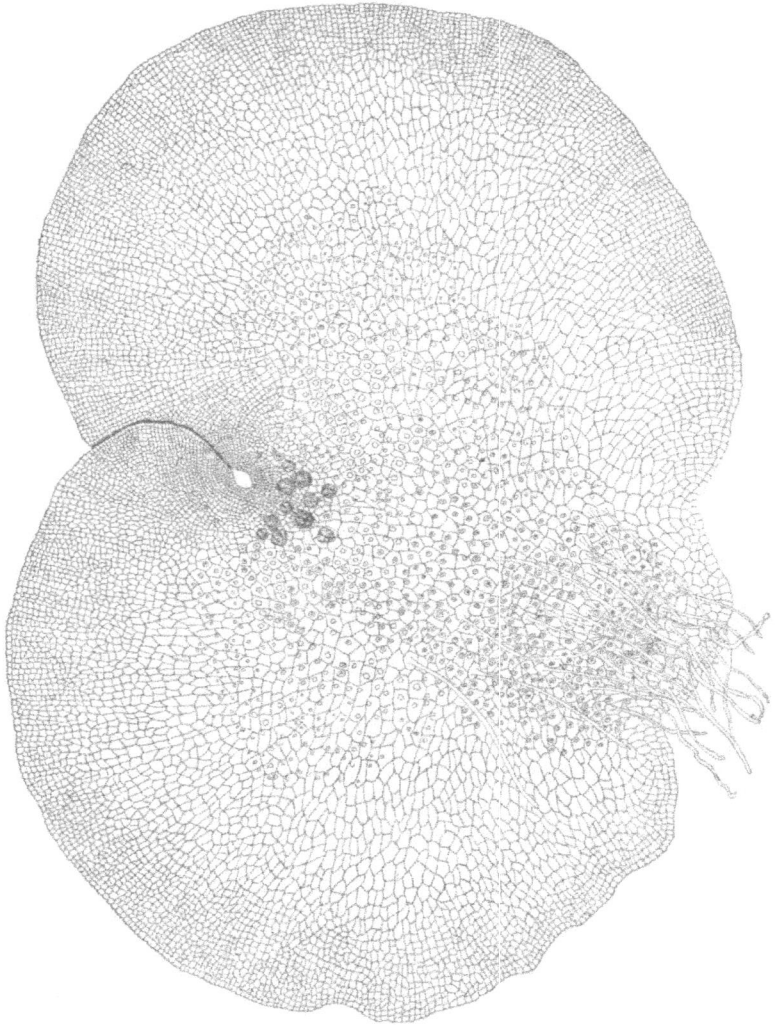

Fig. 13.   Ripe prothallium of Adiantum cuneatum; archegonia in a cluster near the sinus, antheridia scattered over a large part of the surface and among the rhizoids.

the apical cell, several quadrangular cells in a crescentic row
participate. The breadth of the prothallium is increased by
dividing walls perpendicular to the surface of the prothallium,
but parallel with its axis ; while depth is produced by walls also
perpendicular to the surface, but transverse to the axis, cutting
off basal cells, each of which then divides by a wall whose plane
is parallel with the surface of the prothallium. The lower of
these cells is usually the larger. By a perpendicular wall it is
divided into two cells, one of which may become the mother

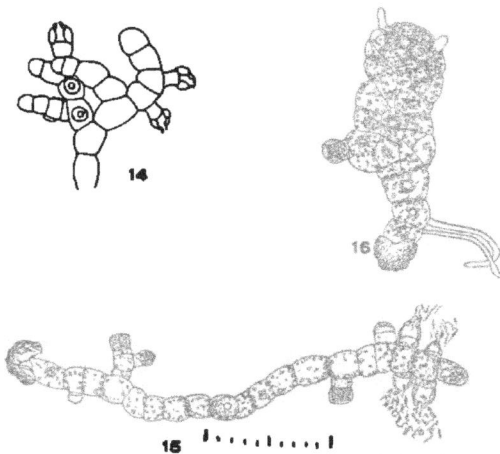

Figs. 14, 15, and 16. Various forms of male prothallia of Niphobolus crispum corymbiferum. In Fig. 16 are two trichomes, at the dystal end sometimes developed on the
margin of various prothallia. Magnified 6 times more than the scale; scale = 1 mm.

cell of an archegonium. By the more rapid increase of these
lower cells the cushion projects below the surface of the prothallium.

Even full-grown prothallia are quite small, perhaps averaging
from 4 mm.–6 mm. Sometimes they are much larger, quite frequently twice this size. Goebel records one of *Osmunda*, which
had grown several years, to be 4 cm. long. In colour they vary
from light yellowish green to dark green.

The sexual organs may be produced both on the same prothallium, when the latter is said to be monœcious. This is
probably the more common. They are, however, produced on
separate prothallia, which are then diœcious. Sometimes a
prothallium bears antheridia for a period, and then archegonia
appear ; *i.e.* it is proterandrous. Very commonly a few antheridia appear before the archegonia. In some species male prothallia are quite small, while the hermaphrodites are large.
This dimorphism of the prothallium has been long known, and
is probably largely due to conditions of nutrition.

The form and size of the prothallia are greatly influenced by
the amount of nutriment, as well as by the more or less crowded
condition. When the spores are sown
thickly and under such conditions that
many prothallia obtain but little nutriment, they may be very rudimentary
and consist of nothing but simple or
variously branched protonemal threads.
The normal prothallium in the Polypodiaceæ, when not crowded, is flat and
quite regularly cordate, or the wings
may be somewhat rectangular, while the
margins are more or less wavy or fluted.
When they are crowded there is a greater
or less tendency, according to the proximity of the individuals, to grow erect
and to be curved and distorted.

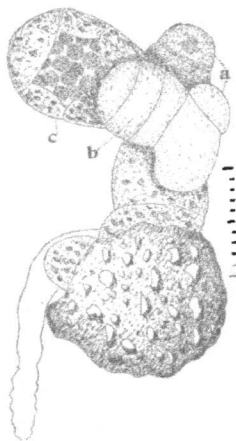

Fig. 17. Very rudimentary
male prothallium of Ni-
phobolus crispum corym-
biferum. *a*, very young
antheridia ; *b*, somewhat
older one, showing the
annular cells; *c*, mature
antheridium. Magnified
30 times more than the
scale; scale = 1 mm.

In *Adiantum cuneatum* the writer
has observed some curious forms of
starved prothallia. In one case, first
noted by a student, the protonemal
thread forked a short distance from the
spore, and the branches extended at
right angles to the primary thread like the arms of the letter
T, each arm bearing a prothallium. In another case observed

the young prothallium produced nearly colourless protonemal threads from marginal cells. Each of these threads bore a prothallium, and in turn produced marginal threads bearing other prothallia. Sometimes even in the Polypodiaceæ there is no primary protonemal thread, but the expanded plate of cells arises directly from the spore. The writer has noted several cases in studying the germination of spores of *Niphobolus crispum*, and in one case two such prothallia were developed from a single spore. In another example the ruptured exospore showed a group of four cells, one of which was producing a rhizoid.

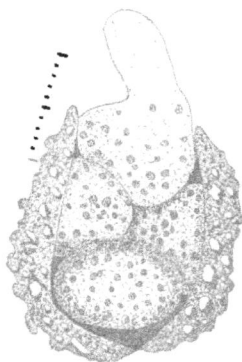

Fig. 18. Spore of Niphobolus crispum corymbiferum divided into four cells before issuing from the exospore; a rhizoid is developing from one cell. Magnified 30 times more than the scale; scale = 1 mm.

Occasionally also in the Polypodiaceæ as well as in the Cyatheaceæ, according to Bauke, and in the Gleicheniaceæ, according to Rauwenhoff, a cell mass is developed directly from the spore.

In the Osmundaceæ the prothallium is usually ribbon-like, with a midrib of several layers of cells, along the edge of which the archegonia are developed. As a rule it arises directly from the spore without the intervention of the protonemal thread. This more simple condition of the prothallium in *Osmunda* frequently closely resembles the thallus of the Hepaticeæ. Campbell has found that the prothallium of *O. claytoniana* is frequently more expanded and resembles that of the Polypodiaceæ.

The growing point of the prothallium is sometimes lateral in the Polypodiaceæ, in which case the prothallium is not cordate. This is very common in the Schizæaceæ.

Goebel found that the prothallium of *Gymnogramme leptophylla* is very irregularly lobed, there being several growing points on the margin. A peculiar outgrowth from the pro-

thallium forces its way into the soil, where it develops into
a tuber-like organ rich in reserve material, and called the
fertile shoot.   The antheridia and archegonia are borne on
its upper side.   If fertilization does not occur, two new prothallia grow out from the tuber.   As the prothallia die away,
adventitious shoots usually grow from the margin or surface.
Many of these are tuber-like, but differ from those of the
primary fertile shoot in bearing only antheridia.   This stage
of *Gymnogramme leptophylla* is therefore perennial, the tubers,
after a period of rest, during which they resist drying and
other changes, developing new prothallia.   Its sporophyte is
annual.   Marginal shoots also occur on the prothallia of other
ferns.

In the Hymenophyllaceæ the prothallium is very variable.
Frequently it consists entirely of a long, much-branched, protonemal growth, bearing antheridia laterally and archegonia
terminally.   In *Trichomanes pyxidiferum* the archegonia are
borne on stalked multicellular bodies, which are lateral outgrowths of the protonema, and are termed archegoniophores.
In other species expanded plates of cells occur along with
the protonemal threads.

Archegonia. — The archegonia are flask-shaped organs, possessing a broad venter, which is sunk in the tissue of the
prothallium.   The neck projects beyond the surface and consists of four rows of cells surrounding a canal.   Each archegonium contains an egg cell (oosphere), a ventral canal cell,
and one or more canal cells.   They arise from superficial cells
of the cushion of tissue at the anterior end of the prothallium.
This cushion of tissue is more richly supplied with protoplasm
than the part of the prothallium on which the antheridia are
borne.   This is in accordance with a general law which obtains
in relation to the orientation of the sexual organs as governed
by the supply of nutriment to different prothallia or to different
parts of the same prothallium.   The office of the archegonium
being the development and protection of the egg cell, the

provision of a canal for the direct approach of the spermatozoids, and a substance to entangle them at the mouth of the
canal, together with the necessity for a supply of nutriment
to the embryo from the prothallium until it shall have gained
a foot-hold in the soil, necessitates a better provision of
nutriment for the archegonia than for the short-lived antheridia.

Usually the archegonia are developed only on the under side
of the prothallium. In two cases the writer found them on
both surfaces on prothallia of *Pteris serrulata*. The prothallia
were crowded so that they stood nearly perpendicular and the
light reached both surfaces. Campbell notes archegonia on
both surfaces of an undetermined fern prothallium. Heinricher
and others have found that lighting prothallia from below
causes the development of archegonia on the upper surface.

The superficial cell, which is the mother cell of the archegonium, first divides by periclinal walls into three cells, the basal
cell, central cell, and neck cell. The central cell is very rich in
granular protoplasm, and has a larger, more distinct nucleus
than the other cells. The basal cell, with other cells of the
prothallium bounding the central cell, develop into the venter
of the archegonium, which consists of smaller cells than those
of the surrounding tissue of the prothallium. The neck cell
is the mother cell of the neck of the archegonium. It divides
by two successive anticlinal walls into four cells, which, by
growth nearly perpendicular to the surface of the prothallium,
produce four rows of cells forming the neck of the archegonium. The two rows of cells on the anterior side of the neck
grow faster than those on the posterior side, and are larger,
sometimes also being greater in number, from four to five on
the posterior side, and five to six on the anterior. This causes
the neck of the archegonium to be curved toward the posterior
end of the prothallium. The central cell, meanwhile, is growing.
It becomes strongly convex on the outer side, and a protrusion
extends between the four neck cells. It now divides by a
periclinal wall into two unequal cells. The smaller one in the

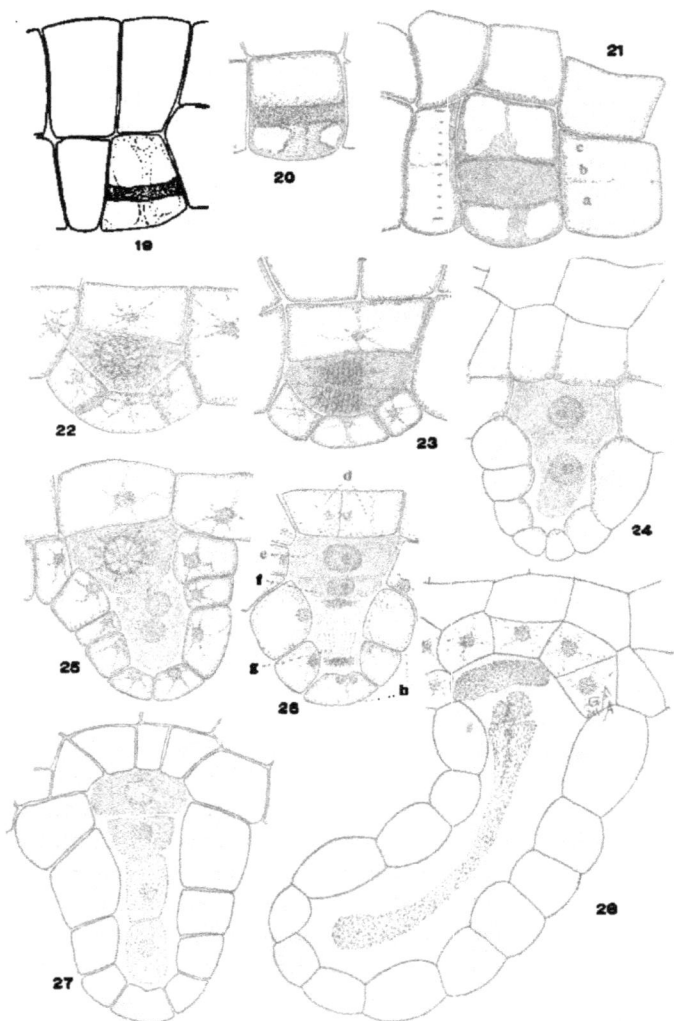

Figs. 19 to 28. Various stages in the development of the archegonium : 21, from Pteris
serrulata; 26, from Adiantum concinnum; the remainder from Adiantum cuneatum.
Figs. 19, 20, and 21: successive stages of the development of the superficial cell of
the prothallium into the three cells: *a*, neck cell; *b*, central cell; *c*, basal cell. In Fig.
26, *e*, the egg; *f*, the ventral canal cell; *g*, the neck canal cell, shows the nuclear
spindle preparatory to the division into two cells. Fig. 27 shows the complete
division of the neck canal cell into two cells. In 28, the canal cells are deliquescing
into the slimy substance. Magnified 30 times more than the scale; scale = 1 mm.
Fig. 24 from preparation by S. G. Harris.

canal elongates by growth, and sometimes at least divides, forming two neck canal cells. The larger cell divides again, and a second small cell is cut off from it, forming the ventral canal cell. The larger cell of these two is the egg cell. At maturity, the ventral and neck canal cells swell, and deliquesce into a slimy substance. The four stigmatic cells at the end of the neck spread wide apart, and permit the slimy substance to protrude part way out of the neck, where it serves to entangle the passing spermatozoids.

The neck canal cells begin so soon after maturity to change to the slimy substance, it is difficult to demonstrate the division into two cells, though it is quite easy to demonstrate two nuclei in the protoplasm of the neck canal cells. In several cases, I have observed an undoubted division into two cells. In another case, the preparation presented a beautiful case of the karyokinetic condition of the nucleus of the canal cell (Fig. 26).

Antheridia. — The antheridia are papillate outgrowths of the prothallium. Each consists of a single layer of chlorophyll-bearing cells forming the wall. This wall encloses at first a single cell rich in protoplasm, which eventually develops into the spermatozoids. On well-developed prothallia, the usual position for the antheridia is among the rhizoids near the posterior end, and extending from this point over a surface outward on each side of the cushion of tissue which bears the archegonia. Especially on old prothallia are the antheridia very numerous on the under surface of the wings each side of the group of archegonia, but usually some distance from the margin of the prothallium. In two cases, where prothallia of *Pteris serrulata* were so crowded that they stood erect, permitting light to reach both sides, antheridia were found on both surfaces. In quite young prothallia, they are frequently borne on the margin. Many male prothallia are very simple, consisting only of a protonemal thread with well-developed antheridia extending as branches in various directions. In some cases, they are very rudimentary, the protonemal threads of one or a

very few cells maturing, with several antheridia clustered at the
end.   The protonemal thread may branch several times irregularly, bearing antheridia on all the branches.   All stages, from
very simple rudimentary male prothallia, which recall the male

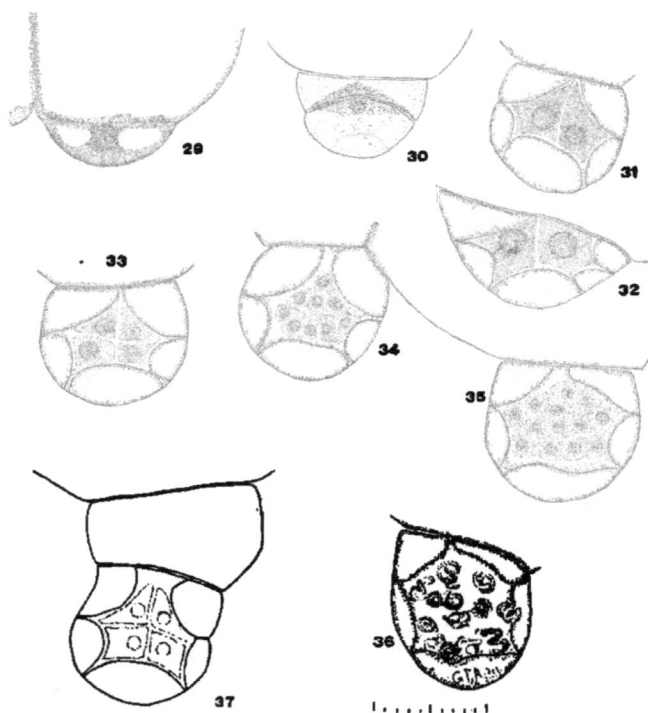

Figs. 29 to 37.   Various stages in the development of the antheridia: 32 and 37 from
    Pteris serrulata; the remainder from Adiantum cuneatum.   Various stages in the
    development of the central cell to form the mother cells of the spermatozoids are also
    shown.   In Fig. 36, the spermatozoids are mature.   In Fig. 37, the antheridium
    possesses a stalk cell.   Magnified 30 times more than the scale; scale = 1 mm.
    Figs. 34 and 35 from preparation by K. M. Wiegand.

prothallia of the Salviniaceæ, through various conditions of
protonemal development, to the more or less expanded plates
of cells, are found.

The antheridium begins as a small protuberance of a superficial cell of the prothallium, so that it is difficult at first to tell whether an antheridium or a rhizoid is to result. But very soon the young antheridium shows a richer and more granular content of protoplasm. Growth of the protuberance continues, and finally it is delimited from the cell of the prothallium by a periclinal wall. The next wall is shaped like a funnel, or watchglass, more rarely straight, the point, or convexity, usually reaching the first, or basal, wall. The third wall is nearly like that of a half-sphere, its convexity lying parallel with the outer convex wall of the antheridium, its rim meeting the preceding wall, thus forming a central cell. This central cell is very rich in protoplasm, and possesses a very distinct nucleus about which granular protoplasm is arranged in an irregularly stellate fashion. A fourth wall now arises similar to the second, cutting off from the end of the antheridium the cap cell. The central cell is the mother cell of the spermatozoids. Sometimes a short stalk cell is developed before the antheridium begins to differentiate.

The superficial cells contain chlorophyll, and they constitute the wall of the antheridium. Two of them are annular, i.e. in the form of complete rings, surrounding the central cell. Because of the thinness of their walls near the point of the transverse diameter of the central cell, in optical section it not infrequently appears as if each annular cell were composed of several cells.

The central cell by division develops into the mother cells of the spermatozoids. It first divides by a longitudinal wall into two cells. The second wall is also longitudinal and perpendicular to the first, thus forming four cells arranged as quadrants when seen from the end of the antheridium. Division proceeds until quite a large and variable number of mother cells results.

Spermatozoids. — The mature spermatozoids are spirally coiled bodies in the form of a screw, with numerous cilia attached to the anterior smaller end. The spermatozoid is developed prin-

cipally from the nucleus of the mother cell. The granules of
the nucleus finally become arranged in the form of a spiral.
The anterior end, according to Strasburger, is developed from
the cytoplasm, and the cilia are outgrowths of this part. The
spiral is smaller at the anterior end, and before motion begins

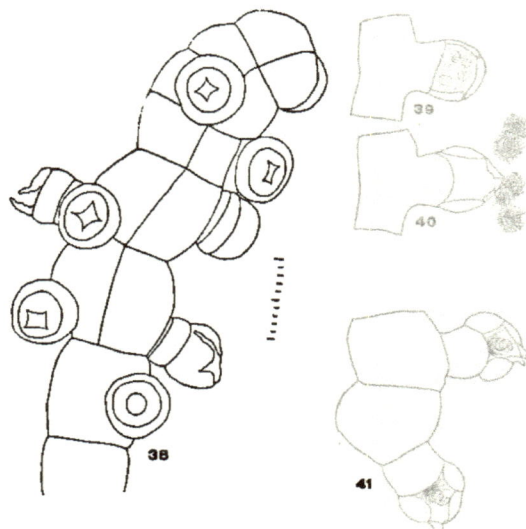

Figs. 38 to 41. Niphobolus crispum corymbterum. Fig. 38, distal portion of male prothallium, showing different views of the antheridia, some open; 39 and 40, showing
the change in the form of the superficial cells at time of the expulsion of the spermatozoids, both figures being drawn from the same specimen. Fig. 41, two dehiscent
antheridia, each containing a spermatozoid rotating in the base. Magnified 30 times
more than the scale; scale = 1 mm.

the cilia are coiled about it. The body of the screw is somewhat flattened, being usually broader at the posterior end.

When the spermatozoids are mature, a portion of the cytoplasm changes to a mucilaginous substance, so that they are
partly freed. At this stage also the superficial cells of the
antheridium possess a strong avidity for water. Thus, when

water comes in contact with them, during rain, or when watered
artificially, these superficial cells absorb quantities of it, which
produces such a pressure upon the contents of the antheridium
that it becomes ruptured.   The rift is irregular, and usually
takes place in the cap cell.   By this means the spermatozoids
are forcibly and usually completely expelled from the antheridium.   For a few moments after expulsion they remain nearly
motionless.   Slight voluntary movements sometimes occur
before the rupture of the antheridium.   After a few moments
of comparative quiet they frequently seem to awake to life
suddenly.   For usually from a state in which only a slight
movement of the cilia is perceptible, suddenly a violent whirling
of the lashes begins, and the spermato-
zoid darts away, dragging behind it for
a time a hyaline sphere, a remnant of
the mother cell.   When no mature
archegonia are near, the spermatozoid
continues in an irregular course, darting
hither and thither for fifteen to thirty
minutes or more.   Finally the movement becomes slower and slower, until
with slow and feeble lashes the cilia are
not able to move the body for any considerable distance, and
eventually all voluntary movement ceases.   When mature
archegonia are near, the spermatozoids in passing are caught
in the slimy substance which is expelled at the mouth and held
there in great numbers.   Movement continues, but the activity
is impeded by the slime, and the spermatozoids plough their
way down the funnel-shaped mouth toward the neck.   In
dehydrating prothallia and infiltrating them with some substance preparatory to sectioning, great numbers of spermatozoids are sometimes retained in the slime at the mouth of the
archegonium in section.

The action of the superficial cells of the antheridium in the
expulsion of the spermatozoids probably assures the escape of

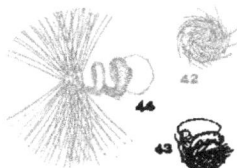

Figs. 42 to 44.   Different views
of spermatozoids in a quiet
condition; Fig. 44. in motion; from Adiantum con-
cinnum.

more spermatozoids than would take place were they left to escape of their own movement after rupture of the lid cell takes place.   In two cases which I observed a spermatozoid was left in an antheridium.   In each case the anterior end happened to be directed toward the base of the antheridium. When movement began, the spermatozoid madly ploughed for more than half an hour against the base of the antheridium, and did not make its escape.

In *Osmunda*, according to Campbell, the mother cell of the antheridium is divided by a wall in such a way that a triangular apical cell is produced, from which several successive tabular cells are cut off.   Eventually a convex wall parallel with the convex outer surface of the apical cell arises, which cuts off a central cell, the primary mother cell of the spermatozoids.

Fertilization. — If a prothallium with mature antheridia and archegonia be inverted on a glass slip and mounted in water for microscopic observation of the under surface, after a few moments, five to twenty minutes, it will be noticed that the spermatozoids are being expelled from the antheridium.   In a few moments more they commence to swarm by whirling rapidly about.   By chance a great many of them pass into the region of the archegonia, and if the attention be now fixed on this spot, hundreds of spermatozoids will be seen swarming in and out among the projecting necks of the archegonia. It will also be noted that the necks of certain archegonia have opened by the wide spreading apart of their four stigmatic cells, and that a slimy, granular, stringy mass, the disorganized canal cells, is protruding.   In this many of the spermatozoids become entangled.   Some free themselves and swim away, but many of them do not escape.   The slimy substance of the canal cells appears to be arranged in numbers of strings which lie parallel with its walls, and at the mouth radiate therefrom. In this the spermatozoids can be seen, the smaller end of the coil directed toward the canal.   Some can be seen quite far

down in the canal, but at the open end they are sometimes
crowded very close together.   Most prothallia are unsuitable
for observing the entrance of the spermatozoid into the egg.
After numerous sections Strasburger succeeded in observing
it in *Pteris serrulata.*   However, the prothallia of *Ceratopteris
thalictroides* were found to be excellent objects to employ for
this purpose.   Because of their thinness and their translucency

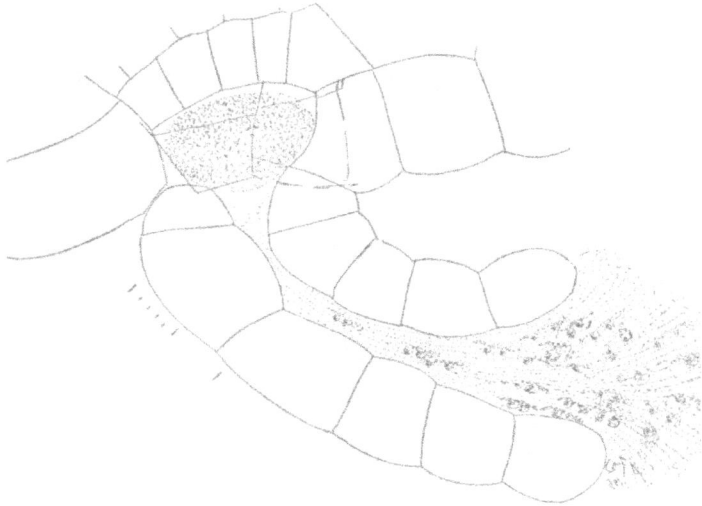

Fig. 45. Dehiscent archegonium of Adiantum cuneatum, var. princeps, showing spermatozoids making their way down through the slime in the neck of the archegonium.
Magnified 30 times more than the scale; scale = 1 mm.

as well as the position of many of the archegonia near the
margin, he was able to see the egg and the entire canal in
optical section.   According to him several spermatozoids may
enter the cavity at the ventral surface of the central cell.   In
passing through the slime in the canal the spermatozoids lose
the vesicle.   They rotate more slowly on their axis, and the
coil becomes longer and more open.   Upon reaching the cavity

at the ventral surface of the egg, the coil partly closes up again and the rotation becomes more active. After a time one is so situated that its anterior end comes in contact with the receptive spot of the egg, a light spot in the protoplasm at the ventral surface. It becomes fixed unless soon crowded away by the jostling of the other spermatozoids. It continues to rotate on its axis and slowly sinks within the egg, and in a few minutes becomes invisible. Campbell was able to see the entrance of the spermatozoids into the cavity of the central cell in living specimens of *Osmunda*, the position of some of the archegonia favouring the observation. The spermatozoid fuses with the nucleus of the egg and fertilization is accomplished.

Without the influence which the addition of this new substance, the body of the spermatozoid, exerts upon the egg, it could not develop into a new plant. Probably the majority of cases where fertilization does take place, the fertilized egg fails to develop into the embryo, for it is a common thing for several eggs to be fertilized on a single prothallium, but the large majority of prothallia perish without the development of a new plant.

II.

EMBRYO.

AFTER fertilization the egg increases in size and completely fills the space in the base of the archegonium. Sometimes it

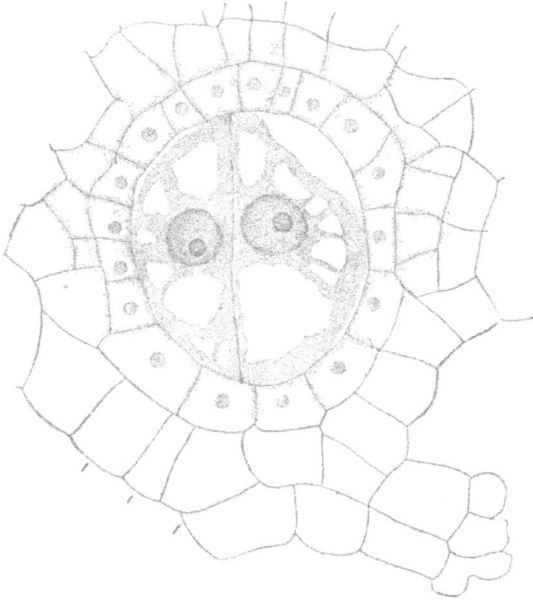

Fig. 46. Two-celled embryo of Pteris serrulata, surrounded by the venter of the archegonium and prothalline tissue. A remnant of the neck of the archegonium is still attached. The basal wall has divided the embryo into two unequal segments, the anterior or stem segment being smaller than the posterior or root segment. Magnified 30 times more than the scale; scale = 1 mm.

23

also arches out with a stronger convexity at the junction of the
canal. In experiencing this change the embryo becomes oval,
the form being determined by that of the cavity. The longer
axis is perpendicular to the surface of the prothallium. The
embryo next experiences growth the influence of which, so far
as cell multiplication is con-
cerned, is in a direction parallel
with the axis of the prothallium. The first division wall
of the young embryo is nearly
or quite parallel with the axis
of the archegonium, conse-
quently nearly perpendicular
to the surface of the prothallium, and it is also perpendicular to the antero-posterior
axis of the prothallium. It
divides the embryo into two
parts, an anterior part and a
posterior part. This first wall
is termed the basal wall. It
separates the stem segment
(anterior segment) from the
root segment (posterior segment). Two other walls follow,
though the order of their succession is not definitely known.

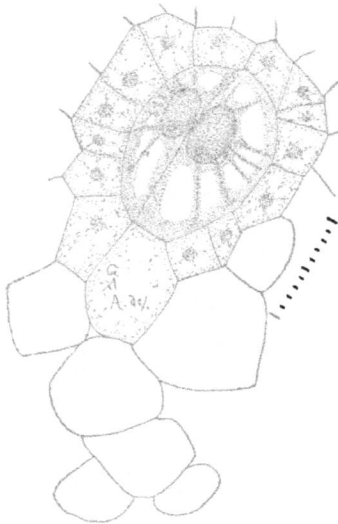

Fig. 47. Two-celled embryo of Adiantum
cuneatum. The stem segment here is
also smaller than the root segment.
Magnified 30 times more than the scale;
scale = 1 mm.

One, the transversal wall, is perpendicular to the basal wall
and parallel with the surface of the prothallium, while the
other, the median wall, is perpendicular both to the basal wall
and to the surface of the prothallium. The eight parts into
which these three walls divide the embryo are termed the
octants, and the four parts formed by the intersection of the
basal and transversal walls are termed the quadrants.

The upper anterior quadrant is the stem quadrant ; the lower

anterior quadrant, the leaf quadrant ; the lower posterior quadrant, the root quadrant ; while the upper posterior quadrant
develops an organ of attachment to the prothallium termed,
inappropriately, the foot.   The stem, root, and leaf quadrants
are so named because those parts of the plant are developed
respectively from those quadrants or parts of the embryo.   Of
the two anterior upper octants only one develops the stem, the
other undergoes little or no farther differentiation.   So of the
two lower posterior quadrants only one takes part in the development of the root.   The stem and root octants lie on the same
side of the median wall, but diagonally opposite as regards the
basal wall.   One of the lower anterior octants is concerned in
the development of the first leaf, or cotyledon.   Both of the
upper posterior octants are concerned in the development of
the foot.

Following the completion of these three walls are two others.
They extend nearly parallel with the basal wall : one, the hypobasal wall, across the posterior part ; the other, the epibasal
wall, across the anterior part.   These walls are slightly convex
on their inner faces, *i.e.*, on the side toward the basal wall.
The stem and root octants serve as tetrahedral apical cells, and
by successive walls nearly parallel with the basal, transversal,
and median walls, the form of the apical cell is preserved.   According to Goebel, in the leaf quadrant no walls arise parallel
with the transversal wall.   In *Osmunda* Campbell found the
leaf octant to serve as the three-sided apical cell, and it is quite
probable that the same thing occurs in other ferns whose leaf
possesses a three-sided apical cell.   Very early in the development of the root segment, walls arise parallel also to the outer
surface of the apical cell, the first cell so cut off being the
beginning of the root cap.

In Fig. 48, a longitudinal section of a young embryo of
Adiantum concinnum, it can be seen that in the tetrahedral root
segment, besides the hypobasal wall nearly parallel with the
basal wall, one wall has formed nearly parallel with the trans-

versal wall, and the nucleus of the apical cell shows the nuclear
spindle in process of segmentation prior to the formation of

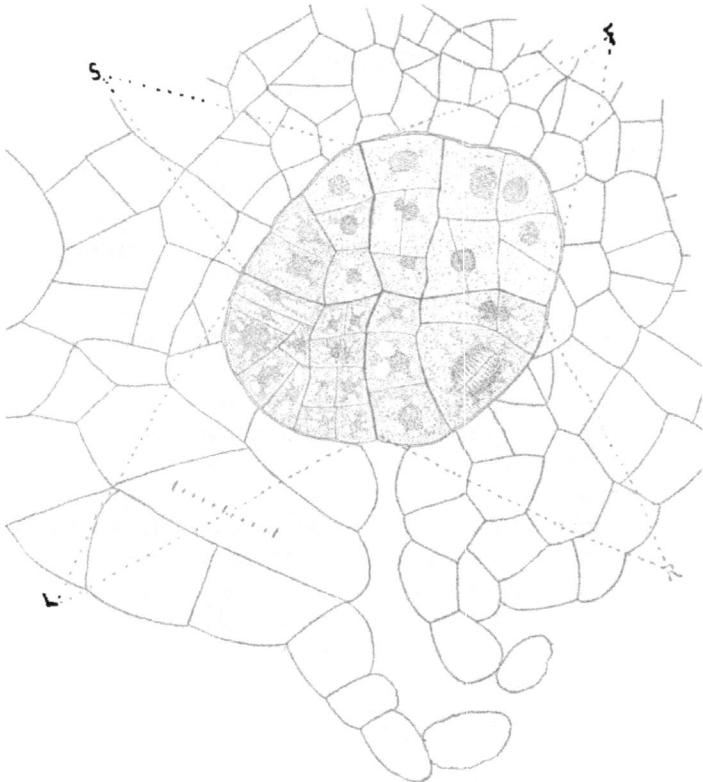

Fig. 48. Young embryo of Adiantum concinnum. A remnant of the neck of the archegonium is attached to the calyptra. The basal wall is in a plane continuous with the
canal of the archegonium. The direction in which the neck of the archegonium
points indicates the root, or posterior segment; on the anterior side is the epibasal
wall and on the posterior side is the hypobasal wall, the transversal wall crossing these
nearly in the middle. *L*, leaf quadrant; *S*, stem quadrant; *F*, foot quadrant; *R*, root
quadrant. Magnified 30 times more than the scale; scale = 1 mm.

the wall cutting off the root cap. In the leaf segment, which
appears always to grow more rapidly at first than the root seg-

ment, the apical cell has been formed and several cells have already been separated from it. The stem segment is always belated in growth, and shows little development in the figure. The foot segment, on the other hand, shows greater growth in extent than any of the others. It can also be seen that there is a definite variation in the richness and granular condition of the protoplasm in the different segments. The foot segment is poor in protoplasmic content, while the cells of the prothallium with which it is connected are very rich in protoplasm, and continue to be so throughout the dependence of the embryo upon it. The root segment shows richer and more granular protoplasm, while the stem and leaf segments possess very rich granular protoplasm, that in the leaf segment exceeding that in the stem segment.

This variation in the richness of the protoplasm bears a definite relation to the functions of the four segments. The function of the foot is to convey nutriment from the prothallium to the essential parts of the embryo. It has need, therefore, of a comparatively small amount and relatively poor quality. Its growth is soon limited, since its function is to be performed for a short time only. Although the function of the root is ultimately to collect and transfer nutrient solutions, it requires a long period of growth in order that it may reach the soil, fix the plant, and then extend for a considerable distance through the soil in search of food. The young root segment therefore shows a richer content of protoplasm than the foot. The stem and leaf segments have higher functions to perform, the leaves ultimately bearing reproductive bodies. While the first leaves never attain this end, it is very important that they should be developed early, since they act as assimilatory organs, and elaborate starch from the carbon dioxide of the air. The stem is concerned chiefly in the production of new leaves, and while in its growing point we would look for a richer content of protoplasm than in the root, it is not of such immediate importance as in the leaf.

Turning now to Fig. 49, we see this law of the distribution of the quality of the protoplasm more highly exemplified in an older embryo.   It is also to be noted that the older cells of the

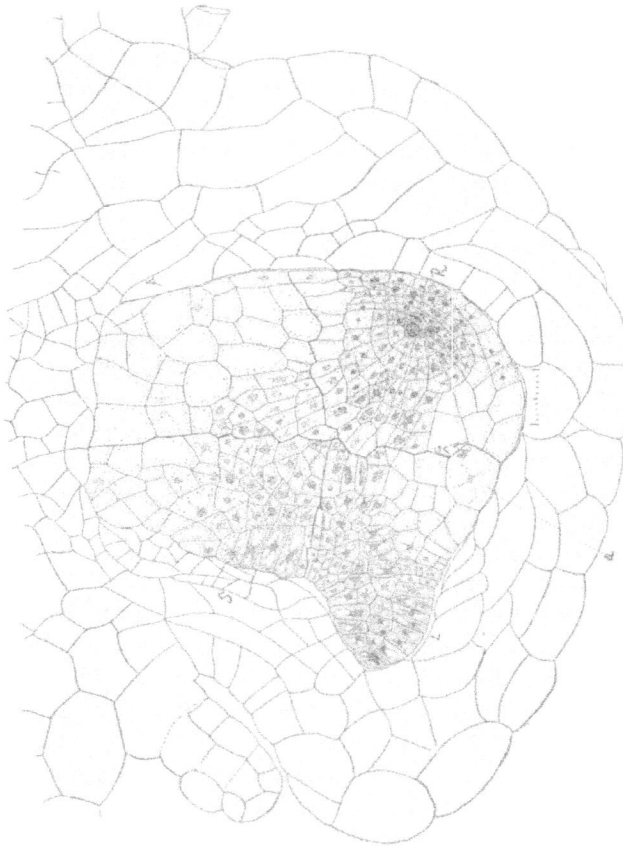

Fig. 49. Young embryo of Adiantum concinnum (considerably older than that in Fig. 48). The calyptra is here very large, and in front of it is shown a section of an archegonium attached to the prothallium. L, leaf; R, root; F, foot; S, stem. Magnified 30 times more than the scale; scale = 1 mm.

stem, leaf, and root now are becoming poor in protoplasm. They are losing the power of growth and cell multiplication, are becoming carriers of nutriment, and are taking on the form of cells of the vascular bundles.   At *a* are cells taking on the character of the cortex. The growing ends of the stem, leaf, and root, however, contain rich, granular protoplasm.

If we now examine still older embryos, we will observe steps in the continuation of this process of change in cell content and cell form accompanying progressive change in function. Fig. 50 represents an older embryo of *Adiantum cuneatum*. Those cells which in Fig. 49 were seen to be losing the richness of their protoplasm and to be elongating in form, have become here changed to scalariform vessels, characteristic of the vascular bundles, while the growing points are still more remote from each other.

In noting these various steps in the development of the embryo it will be seen that as the embryo enlarges, that part of the prothallium which surrounds it grows and bulges downward, producing a prominent convexity at this point on the under side, termed the calyptra. Eventually the embryo breaks through the calyptra, the root fastens it to the soil, and taking on the function of collecting nutriment for the plant, the foot ceases its function, and the prothallium disappears.

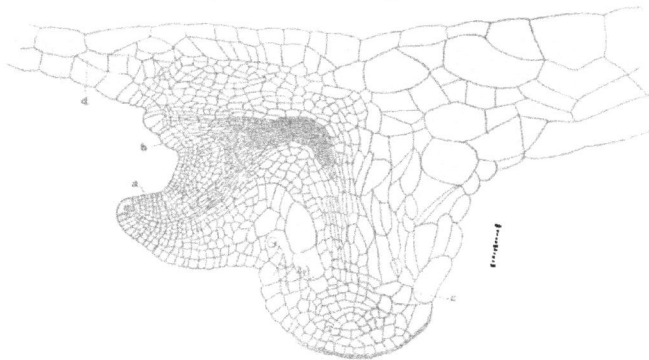

Fig. 50. Embryo of Adiantum cuneatum free from the calyptra, but still attached to the prothallium. *a*, leaf (or cotyledon); *b*, stem; *c*, root; *d*, prothallium. Magnified 5 times more than the scale; scale = 1 mm.

The young fern plant, the sporophyte, having now gained an independent existence by taking firm hold on the substratum, rapidly develops into the fern plant with which we are familiar. The rudiments of the stem, root, and leaf, so nearly alike in the early segmentation, have already diverged quite widely from one another in form, structure, and development, and they are destined to become in the mature plant more widely different.

The young root pierces the soil or substratum on which the plant is growing, or crowds its way into the crevices of rocks, or in the interstices of the bark of trees. Its general course is first of all downwards after having once become well differentiated and developed in the embryo.

On the other hand, the course of the cotyledon and leaves is upward. The tip of the cotyledon soon becomes curved inward toward the prothallium. The general direction of growth is at first forward, *i.e.* towards the anterior end of the prothallium. By the larger size and the more rapid segmentation of the cells on the under side of the young cotyledon, it curves over toward the prothallium. As it approaches the sinus at the anterior end of the prothallium, it ascends and passes up between the open lobes ; or if the lobes overlap, as sometimes happens, it passes out beyond them. This curling inward of the cotyledon, and the more pronounced curling of the older leaves, is characteristic of ferns, and they are said to be circinate in vernation.

Fig. 51. Young plant of Pteris serrulata, with prothallium still attached.

It was once asserted that gravitation controlled the position of the root and other parts of the plant in the young embryo, so that if the prothallium were inclined at various angles, the root and other parts would take the same position relative to the direction of the force of gravitation that they did when the prothallium was perpendicular to the direction of this force. But Leitgeb showed, by growing prothallia in various positions, and by lighting some from below so that the archegonia were produced upon the upper surface, that the parts of the young embryo always had the same position relative to the archegonia that they had when the prothallia were grown horizontally, and lighted from above.

Heinricher, by growing prothallia of *Ceratopteris thalictroides* in such a way as to throw light upon them both from below and above, succeeded in producing archegonia on both surfaces. On two such prothallia, two eggs, one each in an archegonium above and below, became fertilized, and developed into embryos. The parts of the embryo in the archegonia upon the upper surface were in the same position relative to the archegonia that they were below, or in opposite position with reference to the direction of the force of gravitation, or to the upper and lower surfaces of the prothallium.

Several archegonia on a single prothallium may possess eggs which become fertilized, but rarely more than one develops into a perfect plant capable of an independent existence. The cases just cited from Heinricher show that two embryos are sometimes developed, but we have no assurance that both could have gained an independent existence. Rauwenhoff records several cases of two embryos being developed on single prothallia of the Gleicheniaceæ. It is not stated that each can become independent plants. Campbell found two embryos on a single prothallium of *Osmunda*, but he states that one was far in advance of the other, and would probably have starved it out.

In my studies of the Polypodiaceæ, I observed a prothallium of *Adiantum cuneatum* which showed two well-developed cotyledons of apparently about the same age, and growing parallel to each other. Examining the specimen carefully, it was found that both cotyledons were attached to the under surface of the prothallium side by side, and by the base of each was a very young leaf, the tip of which was rolled up in circinate fashion.

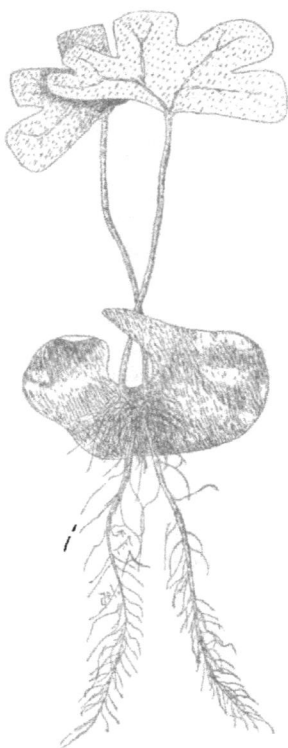

Fig. 52. Adiantum cuneatum; two independent plants from a single prothallium.

Two well-developed roots also issued side by side a little to the rear of the cotyledons. They gave every appearance of being two well-developed plants fixed to the soil, and capable of an independent existence. A sketch was made of the prothallium with its two plants, and is shown in Fig. 52.

It occurred to me that possibly this apparent development of two perfect plants from a single prothallium might be some abnormal condition of a single embryo in which the stem and root, or possibly the stem and root segments, had forked at a very young period in its existence. To be certain what the real condition of things was, the bulk of the cotyledons and roots was cut away, and the prothallium and attached parts were dehydrated, infiltrated with collodion, and cut longitudinally into sections. From the point of starting into one embryo to the passing out of the other, all the sections were preserved and mounted serially. A study showed two perfect plants.

It will now be in order to take up the farther study of the plant parts separately.

III.

MORPHOLOGY AND ANATOMY.

STEM.

Morphology. — In the Polypodiaceæ the stem is usually creeping or ascending, and is termed the rhizome. In *Polypodium vulgare* it creeps upon the surface of the ground. In *Adiantum pedatum* it lies usually in the surface soil, or is covered by leaf mould. In *Pteris aquilina* the rhizome is

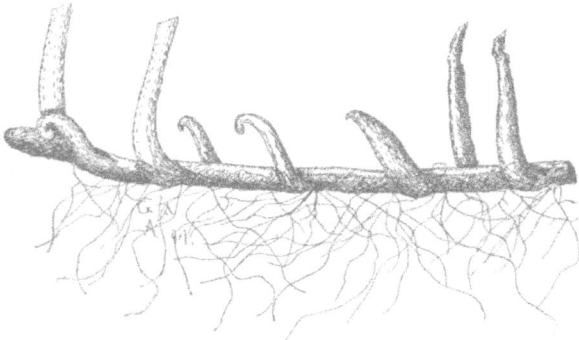

Fig. 53. Stem (rhizome) of Onoclea sensibilis.

several inches, sometimes six to twelve, below the surface, and describes quite a sinuous course. It is somewhat flattened, and two lateral ridges give it a bilateral symmetry. In some species the rhizome creeps over rocks or trees, and in many

33

others it rises obliquely in the soil. The largest stems are found in the Cyatheaceæ, where are found the erect, often columnar, trunks of the tree ferns. Contrasted with these are the delicate filmy plants of many of the Hymenophyllaceæ.

The stem is usually perennial, and the branching is dichotomous. The growing end is usually protected by a profuse development of scales, which converge over the apex. In some species, especially those which rise above the ground, these scales frequently persist over considerable lengths of the stem, and in age are brown in color. The apex is free in *Pteris aquilina, Polypodium vulgare,* and others, while it is concealed beneath the leaf-bud in many.

The growth of the stem proceeds from an apical cell, which is either wedge-shaped, and probably accompanies creeping stems with a bilateral structure, or it is three-sided, and usually accompanies erect or ascending stems with crowded leaves arising on all sides. The successive oblique divisions of the apical cell give a stratified appearance to the resultant tissue,

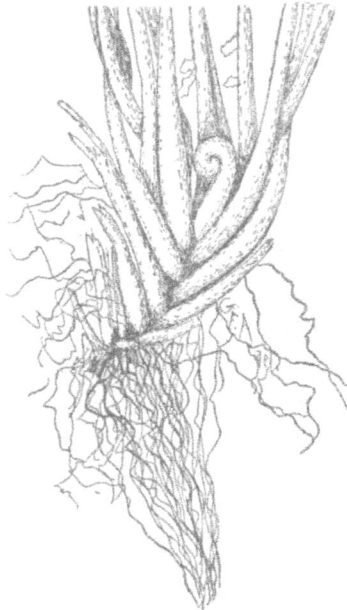

Fig. 54. Stem (rhizome) of Aspidium acrostichoides, traced from a photograph.

which soon disappears by the apparent merging of the elements from repeated cell divisions, and pressure exerted by the development of the leaves.

Anatomy. — The fundamental tissue of many species does not undergo any subsequent change, but remains in the form of thin-walled parenchyma. In *Pteris aquilina* portions of the fundamental tissue change to sclerenchyma. In a transection of the stem of this species near the centre are seen two groups of sclerenchymatous tissue in the form of bars, which form longitudinal ribbon-like strips that run parallel through the stem. The cell walls are dark brown and traversed by numerous lacunæ connecting the cells. In age, groups of cells here and there through the parenchyma change to scleren-

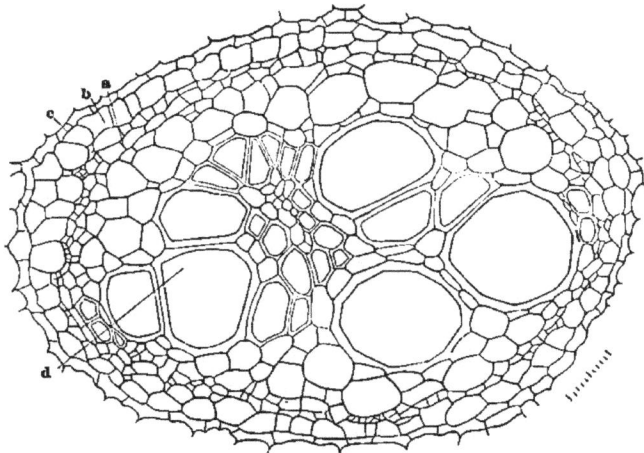

Fig. 55. Transection of vascular bundle in stem of Pteris aquilina. *c*, bundle sheath; *b*, phloem sheath; *a*, phloem portion of bundle; *d*, xylem portion of bundle. The thick-walled cells are tracheides; intermingled with them are thin-walled cells, the xylem parenchyma. Magnified 10 times more than the scale; scale = 1 mm.

chyma and are seen as dark points in the transection. An external layer several cells deep also changes to somewhat thinner walled sclerenchyma, forming what is termed a shell. A similar shell is also found in the tree ferns and others. In *Lygodium palmatum* the entire parenchyma outside the single central vascular bundle becomes transformed into thick-walled sclerenchyma, which accounts for the stiffness of old stems.

Sheaths of sclerenchyma are also said to be formed about the several vascular bundles in the stems of tree ferns and in *Polypodium vaccinifolium*. In *Adiantum pedatum* nearly the entire fundamental parenchyma becomes transformed into brown sclerenchyma; only a thin layer each side the single

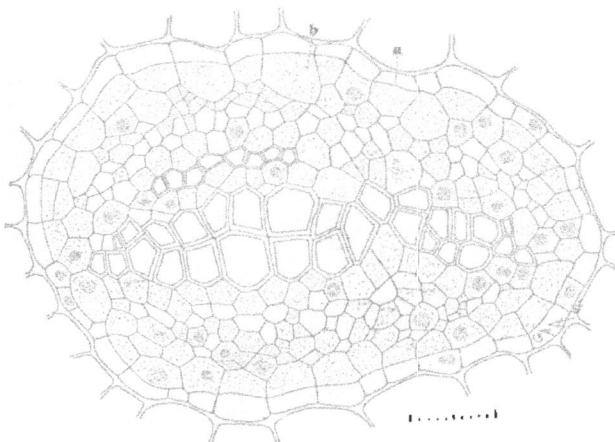

Fig. 56. Transection of vascular bundle in the stem of Polypodium vulgare. *a*, bundle sheath; *b*, phloem sheath. Magnified 30 times more than the scale; scale = 1 mm.

nearly tubular bundle remains unchanged. The medullary sclerenchyma possesses much thicker walls than that in the foliar gaps of the bundle or outside.

The vascular bundles of the stem of most ferns are concentric, the xylem or woody portion being surrounded by the phloem, and the whole encased in a bundle sheath, the endodermis. They are thus closed, growth taking place while the tissues are being differentiated in the procambium about the apical cell. Lying next the bundle sheath inside is a single layer of parenchyma cells, sometimes termed the phloem sheath. The bundle sheath and phloem sheath have a common origin from a single layer of parenchyma cells. The cells of the bundle sheath are narrower in their radial diameter than those

of the phloem sheath, somewhat elongated in a peripheral
line, more early lose their protoplasmic content, and are quite

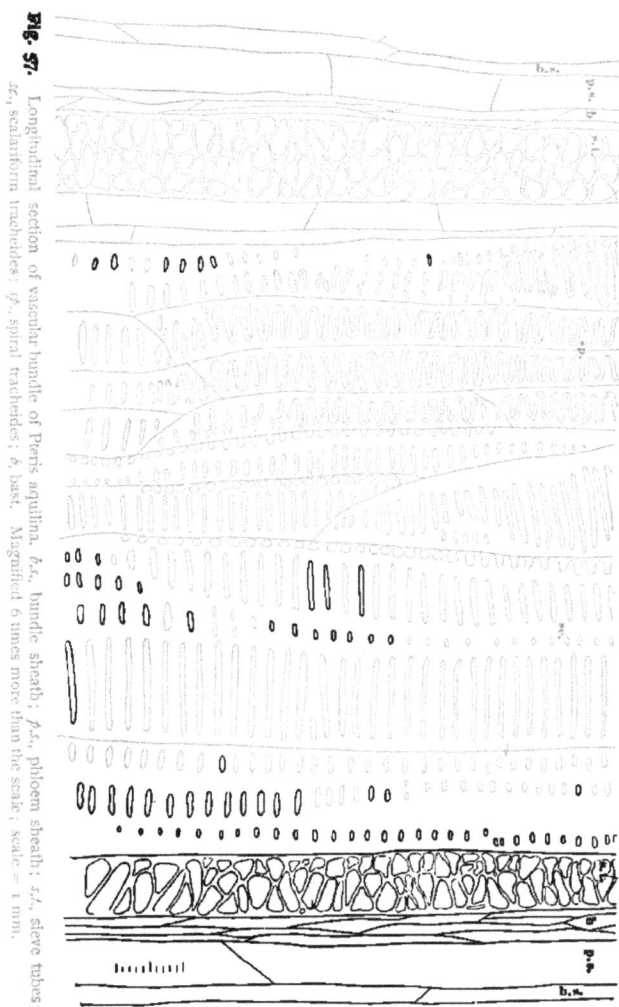

Fig. 37. Longitudinal section of vascular bundle of Pteris aquilina. *b.s.*, bundle sheath; *p.s.*, phloem sheath; *s.t.*, sieve tubes; *sz*, scalariform tracheides; *t.s.*, spiral tracheides; *b*, bast. Magnified 6 times more than the scale; scale = 1 mm.

easily differentiated from the larger parenchyma cells on either
side.   The radial walls of the bundle sheath are usually weaker
than those of the other cells, and in shrinkage, which occurs
in drying, frequently they are broken, and the bundle thus
partly separated from the surrounding parenchyma.

The xylem portion of the bundle is made up of several kinds
of tissue.   Fundamental tissue in the form of rather small,
thin-walled cells is intermingled with the various tracheides,
which possess rather thick walls, variously sculptured.   Scalariform, reticulate, and spiral tracheides occur frequently in the
same bundle, and in addition in *Pteris aquilina* are true scalari-

Figs. 58 and 59.   From Pteris aquilina; 58, longitudinal section of scalariform vessel;
59, longitudinal section of scalariform tracheid.   Magnified 30 times more than the
scale; scale = 1 mm.

form vessels, shown by the perforation of their cross-walls.
The cells of the phloem vary greatly in size, and are intermingled with sieve tubes and parenchyma.

The bundles vary in number, arrangement, and position in
different species, fundamental tissue occupying the intervening
spaces, except, as stated above, where portions become changed
to sclerenchyma.   In *Pteris aquilina* the bundles are in two
groups.   In cross-section of the stem one group, composed
of small oval or irregular bundles, is arranged in the form of
a ring enclosing the bars of sclerenchyma; the other group

is composed of two bundles usually, nearly as broad as the bars of sclerenchyma, parallel with, and inside of, them. In *Polypodium vulgare* several bundles are arranged in a single ring, the medullary parenchyma connecting by radial strips between the bundles with the cortical parenchyma. The same type occurs in *Onoclea sensibilis, Asplenium filix-fæmina,* and

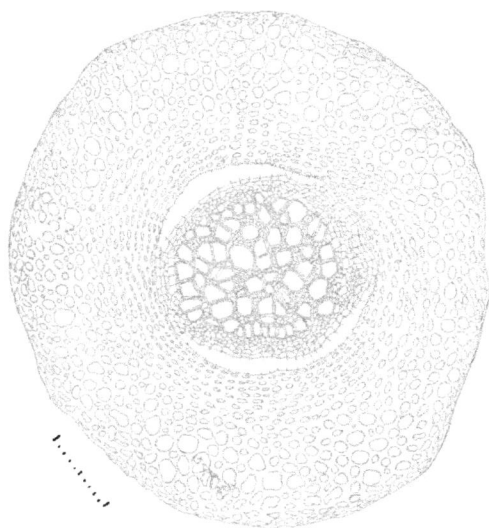

Fig. 60. Transection of the stem of Lygodium palmatum. The radial walls of the bundle sheath are ruptured, thus separating the single central bundle from the sclerosed tissue outside. All of the cortical parenchyma has changed to sclerenchyma. Magnified 6 times more than the scale; scale = 1 mm.

many others, these being cited as examples only, and having been examined by the writer.

In *Adiantum pedatum* a single bundle in the form of a tube with perforations, or gaps, associated with the leaf traces, runs through the stem. At the intersection with the leaf stalk the bundle in transection has the form of a horseshoe, while the bundle of the leaf trace is C-shaped, the open part being next that of the open part of the stem bundle. As the sections pass either side of the centre of the foliar gap, the ends of the horseshoe-shaped bundle approach each other until the margin of the leaf trace is reached, when the bundle in transection forms a complete ring with an oval outline. At the same time the sections about two-fifths the way around show a narrowing of the width of the bundle ring, which proceeds in the series of sections until another foliar gap is reached, when the bundle ring is again open.

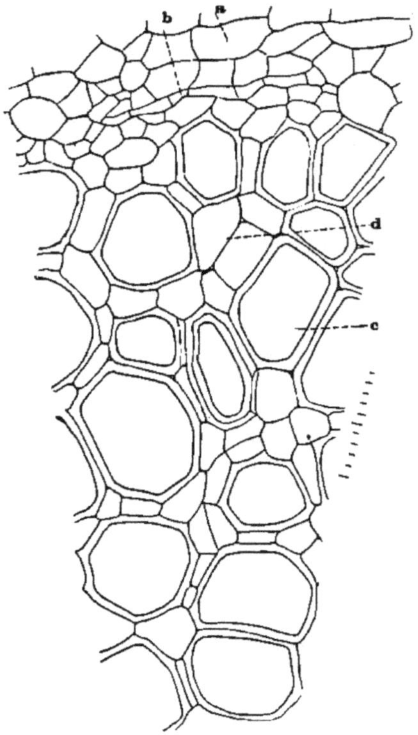

In those ferns which possess several bundles in the stem, the bundles fork and anastomose, forming a cylindrical network. With some of the gaps formed by this network the leaf traces are associated. These gaps usually have a spiral arrangement, frequently showing a definite relation to the phyllotaxy.

Fig. 61. Portion of vascular bundle of Lygodium palmatum. *a*, bundle sheath; *b*, phloem portion; *c*, xylem portion; *d*, xylem parenchyma. Magnified 30 times more than the scale; scale = 1 mm.

In *Lygodium palmatum* and other species of the Schizæaceæ, various species of the Hymenophyllaceæ, and *Gleichenia* of the Gleicheniaceæ, there is but a single central bundle.

In the Osmundaceæ the bundles of the stem are collateral;

i.e. the xylem occupies one side, and the phloem the other. The bundles are arranged in a circle, the xylem occupying the axial side, and the phloem the peripheral side, where it forms a complete ring outside the circle of xylem groups. In *Osmunda* the entire cortical parenchyma becomes trans-

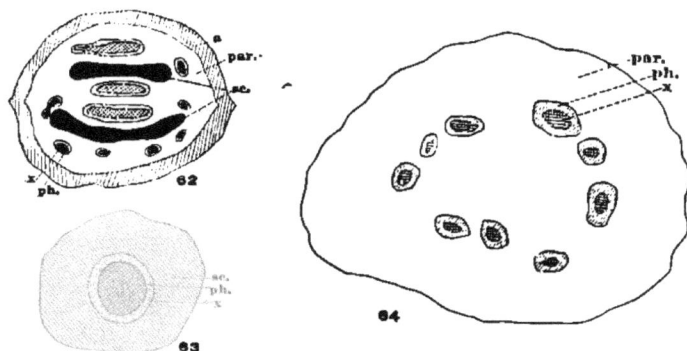

Fig. 62. Diagrammatic section of the stem of Pteris aquilina. *x*, xylem portion of bundle; *ph.*, phloem portion; *sc.*, thick-walled sclerenchyma; *a*, thin-walled sclerenchyma; *par.*, parenchyma. — Fig. 63. Diagrammatic section of the stem of Lygodium palmatum. *x*, xylem portion; *ph.*, phloem portion; *sc.*, sclerenchyma. — Fig. 64. Diagrammatic section of the stem of Polypodium vulgare. *x*, xylem portion of bundle; *ph.*, phloem portion; *par.*, parenchyma.

formed into hard, dark sclerenchyma, through which the vascular bundles of the leaves course upward and outward, several being seen in one transection of the stem. The medullary parenchyma remains unchanged.

ROOT.

Morphology. — The roots arise in succession in the growing end of the stem. They originate in the procambium before the differentiation of the vascular bundles. As the xylem of the bundle is formed, the cells of the radial side of the stem bundle pass obliquely into the bundle of the root, and no gap occurs as in the case of the origin of the leaves.

The growing point of the root is a tetrahedral apical cell, with its faces sometimes slightly curved, especially the side toward the root cap. Triangular tabular cells are cut off usually successively on the four sides. The segments cut off in front develop into the root cap. The lateral triangular tabular segments divide in such a way that the cells formed from the point where they meet in the centre of the root differentiate into the

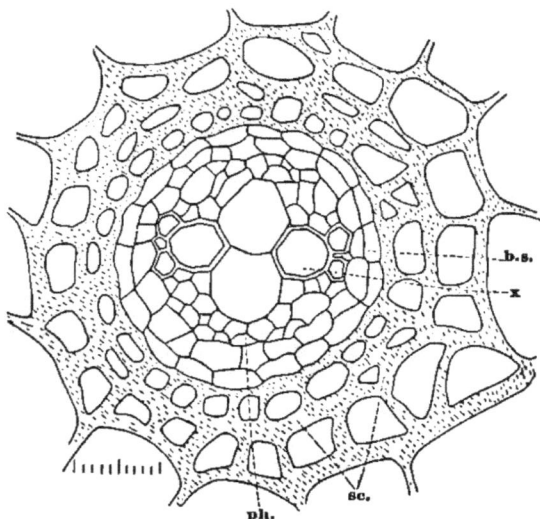

Fig. 65. Section of central portion of the root of Pteris aquilina, showing the single radial bundle and the sclerenchyma immediately surrounding it. *b.s.*, bundle sheath; *x*, one of the two xylem groups; *ph.*, one of the two phloem groups; *sc.*, sclerenchyma. Magnified 30 times more than the scale; scale = 1 mm.

vascular bundles, while the peripheral ones produce the cortex. The young roots grow through the cortex of the stem to the outside. Transections or longisections of the growing end of the stem frequently show the young root with its apical cell and root cap still within the cortex. Roots also arise in some species from the base of the leaf stalk, especially if these are covered with soil. The young roots are covered with root hairs, which are simple non-septate cells.   Similar hairs are also found in some cases on the stems and bases of the leaf stalks.

In tree ferns numerous roots are developed from the aerial part of the stem.   In some cases where the trunk is slender at the base it really appears stouter from the thick mat of roots which extend downward over it.

The roots branch in a monopodial fashion.   Lateral roots are also produced.   They are not connected with the bundle of the root, but arise outside the pericambium from an innermost cell of the cortex (one of the cells of the bundle sheath), which forms a mother cell.   Three obliquely anticlinal walls in this mother cell form the tetrahedral apical cell, and a periclinal wall then forms the primary cell of the root cap.

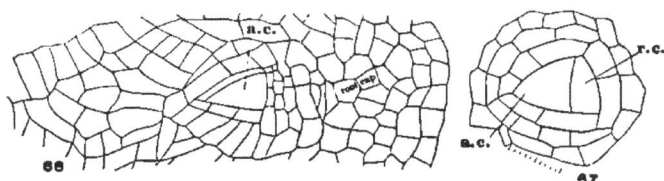

Figs. 66 and 67.   Tetrahedral apical cell of root of Onoclea sensibilis, with surrounding tissue.   *a.c.*, apical cell; *r.c.*, root cap.   Both sections are taken from very young roots still within the tissue of the stem.   Magnified 30 times more than the scale; scale = 1 mm.

Anatomy. — A single central vascular bundle is present, and is surrounded by cortical parenchyma, the innermost layers of which are sometimes sclerosed.   The formation of xylem usually begins at two diametrically opposite points near the outside, the first tracheides formed being narrow and fibrously thickened, and lying side by side.   The transformation then proceeds inward, one or more rows of large scalariform tracheides being formed.   According to Russow, true vessels are only found in the root of *Athyrium* (*Asplenium*) *filix-fœmina*.   This arrangement of the xylem in the root bundle De Bary calls a "diarch" bundle.   According to the same authority triarch and tetrarch bundles are sometimes found in thick roots of some species.   According to Mettenius and Russow, tetrarch and octarch bundles are the rule in species of *Trichomanes*, diarch bundles being rare, while diarch bundles are usual in *Hymenophyllum*.

In *Adiantum pedatum*, which is usually diametrically diarch, I observed one root, the consolidated xylem plates of which formed a crescent, the points of origin of the xylem being nearer on one side, and at the point of consolidation on the other side was an extra large tracheid.

The xylem groups are usually united in the middle by two large tracheides crossing the consolidated plate at right angles. In *Osmunda* and *Todea* of the Osmundaceæ a transection of the root shows the consolidated xylem plates to present an elliptical outline.

The phloem groups alternate with the xylem, giving the appearance of a radial bundle.

SPOROPHYTE.

MORPHOLOGY AND ANATOMY (*continued*).

LEAVES.

Morphology. — The origin of the first leaf in the embryo we have already traced, where it is cut off as a segment from the primary stem segment of the embryo. The other leaves arise by the arching out of a superficial cell, at the side of the growing end of the stem, which becomes the apical cell. In the mature stem at this point of origin of the leaf occurs what is known as the leaf trace, marked by the leaf bundle in the cortex of the stem. In connection with the study of the first leaf, it has already been noted that the leaf grows rapidly in advance of the stem. This is a character of all leaves of ferns, though in many cases the leaves develop very slowly, requiring in the case of *Pteris aquilina* two years. The development is basifugal ; *i.e.* the base is developed first.

The leaves are the most conspicuous part of the plant, and are usually quite well differentiated into two parts, the lamina, and the stipe, as it is called. The function of the stipe is in the main that of support for the lamina, and to continue the channel for the interchange of nutrient matters between different parts of the plant. It is quite stout and hard in *Pteris aquilina;* hard, black, and rather slender in *Adiantum.* In such ferns as *Asplenium* and *Aspidium* it is usually shorter, compared with the extent of the lamina, and in many cases bears numerous brown scales, termed chaff.

The function of the lamina in general is twofold : a physio-
logical one, which includes the phenomena of transpiration,
the reduction of the elements of carbon dioxide, and the production of starch ; and a reproductive one, the development of
spores.   The simplest form occurs in such ferns as *Scolopendrium vulgare* and *Camptosorus rhizophyllus*, the lamina being
normally simple, while in some cases the end forks.   In *Polypodium vulgare* it becomes pinnate; in *Pteris aquilina*, once

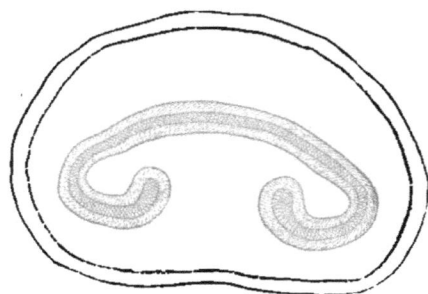

ternate and twice pin-
nate.   In *Adiantum pe-
datum* it is forked into
two recurved divisions,
which bear several sec-
ondary pinnate divisions
on one side.   In various
species of *Asplenium*
and *Aspidium* it is
once, twice, or thrice
pinnate.   The large,
handsome lamina of the
ostrich fern, *Onoclea*

Fig. 68. Diagrammatic section of the stipe of Osmunda regalis, showing position and form of the
single concentric vascular bundle.

struthiopteris, is once pinnate, the pinnæ being pinnately
divided.   These examples will serve as characteristic illustrations of some of the various forms of the lamina.

Anatomy. — The vascular bundles are usually concentric, and
vary in number and form in various genera.   In *Adiantum
pedatum* there is a single bundle which is strongly crescent
shaped or semilunar in transection near the base of the stipe,
while near the junction of the lamina it is much broader in
proportion to its length in the section and comparatively little
curved.   In *Cystopteris bulbifera* there are two bundles, while
in *Pteris aquilina*, in one specimen observed there were ten in
transection, which varied greatly in form and size.   True vessels occur in the stipe of *Pteris aquilina* as well as in the stem.

The vascular bundles which branch from those in the stipe

or rachis, and course into the divisions of the lamina, form the framework. They are termed veins, and either terminate in free ends or in some cases anastomose. The soft parts, or mesophyll, of the leaf is composed of loosely arranged paren-chymatous cells richly filled with chlorophyll, and possessing very large interspaces. Both surfaces of the lamina are cov-ered with an epidermis, the cells of which are well supplied with chlorophyll, usually in much greater abundance than occurs in the higher plants, and is one of the distinguishing features of ferns as com-pared with the higher forms. The epidermis is variously clothed with hairs. Sto-mates of the usual form and with the guard cells charac-teristic of higher plants occur in abundance on the under side of the leaf. They are situated usually in the margin of an epidermal cell. In *Aneimia* and *Polypodium lingua*, Goebel says they are in the middle of an epider-mal cell.

The development of the stomates can be very easily traced by making horizontal sections in the plane of the

Fig. 69. Portion of the epidermis of the leaf of Adiantum pedatum, showing irregular epidermal cells and the stomates in the side of the cells. Magnified 30 times more than the scale; scale = 1 mm.

under side of very young leaves. The epidermal cells are still quite small, and do not present such lobed, interlacing margins as are seen in the fully developed epidermal cells. Their margins are plane or only slightly irregular. The cells are also very rich in protoplasm, with prominent nuclei, and various stages of cell and nuclear division can be seen. A semicircular wall in the margin of an epidermal cell cuts out an oval cell, which sometimes becomes directly the mother cell of the guard cells.   These are formed by longitudinal division of the oval

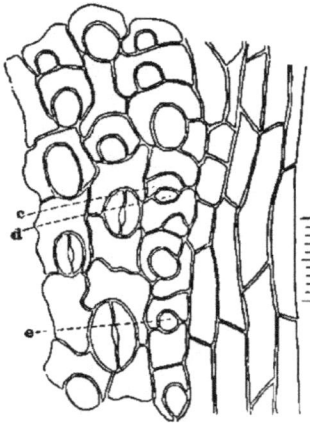

cell.   Sometimes a second semi-lunar wall cuts out a cell from this one, which becomes the mother cell of the guard cells.

Sterile and Fertile Leaves. — The first leaves of young ferns are always sterile.   Of mature plants, if more than one leaf is developed each year, some of the leaves are usually sterile while others are fertile.   Sterility and fertility depend largely upon external conditions, season, soil, habitat, etc.   In some cases, there is no great difference between the sterile and fertile leaves, except that the latter bear spores.   This is quite commonly noted in such ferns as *Polypodium vulgare*, various species of *Asplenium* and *Aspidium*.

Fig. 70.   Portion of very young epidermis of the leaf of Pteris aquilina. *e*, mother cell of a stomate cut off from an epidermal cell by a semicircular wall; *d*, a similar mother cell of a stomate cut off from a cell, which was cut off in like manner from the epidermal cell and prior to it. Magnified 30 times more than the scale; scale = 1 mm.

Dimorphism. — Still in *Asplenium angustifolium*, fertile leaves possess somewhat narrower pinnæ than the sterile ones, and the latter are coarser in general aspect.   In *Aspidium acrostichoides*, the terminal portion of some of the leaves is fertile, while the lower portion is sterile.   Here the fertile pinnæ are much narrower and shorter, presenting a dwarfed appearance.   Yet there are leaves which are entirely sterile.   In a single tuft of leaves from the same plant, it is common to note all gradations of fertility of the various leaves, some bearing only a few fertile pinnæ at the tip.   The fertile leaves of *Pellæa atropurpurea* are simply pinnate at the tip, and usually twice pinnate at the base, and the pinnules are narrowly elongate, while those of the sterile leaf are simple and oval.

In *Onoclea*, the leaves are usually completely differentiated into a sterile leaf and a fertile leaf, presenting a definite case of dimorphism. In the fertile leaf, the pinnæ and pinnules are much dwarfed, and form together a somewhat crowded one-sided spike-like fructification, while the sterile leaves are broadly expanded. Various modifications of this differentiation in *Onoclea sensibilis* occur. For example, leaves are found, many of the pinnæ of which are sterile, while from this condition, various grades of dwarfed fertile pinnæ exist. Underwood calls attention to the fact that probably this latter form of the fertile leaf follows an injury to the plant which destroys the sterile leaf. This would throw the responsibility for the vegetive function solely upon the fertile leaf. In accordance with the law of correlation, then, the increase of the vegetive structure and function in the fertile leaf would lessen the reproductive function, and the partially expanded pinnæ would bear a few sori.

Schizæa pusilla of the Schizæaceæ possesses very slender linear tortuous sterile leaves, the fertile leaves possessing pinnæ arranged in a crowded one-sided arched spike-like fructification.

In *Lygodium*, the leaf resembles a climbing leafy stem. The fertile portions of the leaf are much contracted.

Osmunda presents an interesting variation in the different species. The terminal portion of the leaf is fertile in *O. regalis*, the middle portion in *O. claytoniana*, while in *O. cinnamomea* the leaves are differentiated into a fertile and sterile one. Various modifications of this differentiation occur in the different species. In the fertile portions of the leaf of *O. regalis*, various dwarfed stages of the sterile pinnules occur, in some cases nearly all of the pinnæ being sterile and expanded, a few of them being dwarfed and fertile on the margin, or on both surfaces.

In certain epiphytic ferns, according to Goebel, one kind of leaf is differentiated in species of *Polypodium* to catch and hold humus, and thus serves a function in the supply of nutriment.

Similar leaves were also found in *Bolbophyllum*. In *Platycerium alcicorne*, the staghorn fern of our conservatories, the fertile leaves are narrow and forked. There are also specialized sterile leaves, very broad and oval in outline, which hug very closely around the roots, and serve to catch humus, and conserve the moisture. They are at first green in colour, but in age become brown. They are sometimes called mantle leaves. In New South Wales there are specimens growing around the trunks of standing trees which form massive growths two to three feet in diameter, completely encircling the trunk of the tree.

Fructification. — The fructification of ferns consists of sporangia, containing the spores, usually grouped at definite places on the under side of the leaf, or on the leaf margin. These groups, or fruit dots, technically sori, are oval, circular, linear, or very much elongated and sinuous. They are either naked, or covered by a plate of thin cells of epidermal origin called the indusium. The sorus is usually upon one side of a vein, or at its apex. The indusium varies greatly in form in different ferns. This, together with the form of the sorus, and its position on the leaf, including the variation of the same, form excellent differential characters in the classification of the genera and species. To illustrate this, several examples may be taken from the Polypodiaceæ.

In *Polypodium* the sori are naked and situated on or at the termination of the veins, in a row near each margin of the pinnæ. Perhaps to partly replace the indusium in *Polypodium vulgare* there are numerous multiseptate hairs among the sporangia, the enlarged free end projecting beyond the sporangia. In *Adiantum* and *Pteris* the sori are at the ends of the veins at the margin of the leaf, where they are covered by the incurved margin, which forms what is called the false indusium. In *Pteris* there is in addition a true indusium on the opposite side of the sorus in the form of a flap. In *Adiantum* the main rib, when present, of the pinnules is marginal, while in *Pteris* it is

central.   In a number of genera the sori are oblong or linear,
and extend along one side of fertile veins with the indusium
attached by one edge.   Thus in *Woodwardia* they are in a
chain-like row each side of the midvein of the pinnæ and

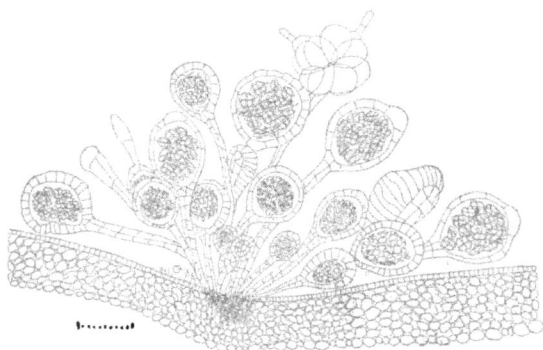

Fig. 71.   Section through sorus of Polypodium vulgare, showing different stages of the
development of the sporangia, and a single multicellular capitate hair.   Magnified 6
times more than the scale; scale = 1 mm.

pinnules.   In *Asplenium* they are separate and extend obliquely
on the leaf.   In *Camptosorus* they are scattered and extend in
various directions, while those nearest the midrib tend to converge in pairs at their outer extremities, forming V-shaped
figures.   In *Scolopendrium* they are closely approximated in
pairs which stand nearly perpendicular to the midrib, appearing
to have a double indusium opening along the middle.

Fig. 72.   Section through sorus and shield-shaped indusium of Aspidium acrostichoides.

Aspidium has a shield-shaped indusium, thus opening all
around.   *Cystopteris* has an ovate one attached by its broad base

somewhat under the roundish sorus. In *Onoclea* there is a hood-shaped indusium fixed by one side and arching over the sorus, and then the closely inrolled pinnule covers both indusium and sorus. In *Woodsia* the indusium is inferior and is attached all around the placenta. It is either open at first, or it covers the sorus, and later opens by irregular, radiating slits. *Dicksonia* possesses an elevated placenta surrounded by a hood-shaped indusium which opens toward the margin on the leaf.

Year by year the old leaves die down, while new ones arise at the growing end of the stem. Frequently in a single season leaves of two or more years may be seen in a single cluster, the old ones dead and in various stages of disorganization.

V.

SPORANGIA.

Development of the Sporangia. — The sporangia are capsules which contain the reproductive bodies known as spores. They vary in form and structure in the different families. Their development as known in the Polypodiaceæ will be described and followed by a comparison of their structure and function in the different families. In the Polypodiaceæ they are said to be trichomic in their origin ; *i.e.* they originate from a single superficial cell.

The cell surface from which they arise is termed the placenta. The cells of the placenta are richer in protoplasm than those of the other epidermal surfaces at this time. The mesophyll beneath is also rich in protoplasm. The placenta is in close communication with a vascular bundle of the leaf, and is thus in direct communication with the channels for the conveyance of nutriment.

A transection through a placenta at the time of the early development of the sporangia will show various degrees of prolongation of individual cells of the surface. The cell. first of all elongates perpendicularly, so that its outer surface is raised above that of its fellows. This elongation continues, and as the growing end of the cell passes the line of the surface it broadens out, because it is relieved from the pressure of adjacent cells and is strongly turgescent. The more granular protoplasm collects in the outer end of this elongated cell and surrounds the prominent nucleus. From the granular proto-

plasm immediately surrounding the nucleus granular strands
radiate in all directions to the periphery of the cell lumen,

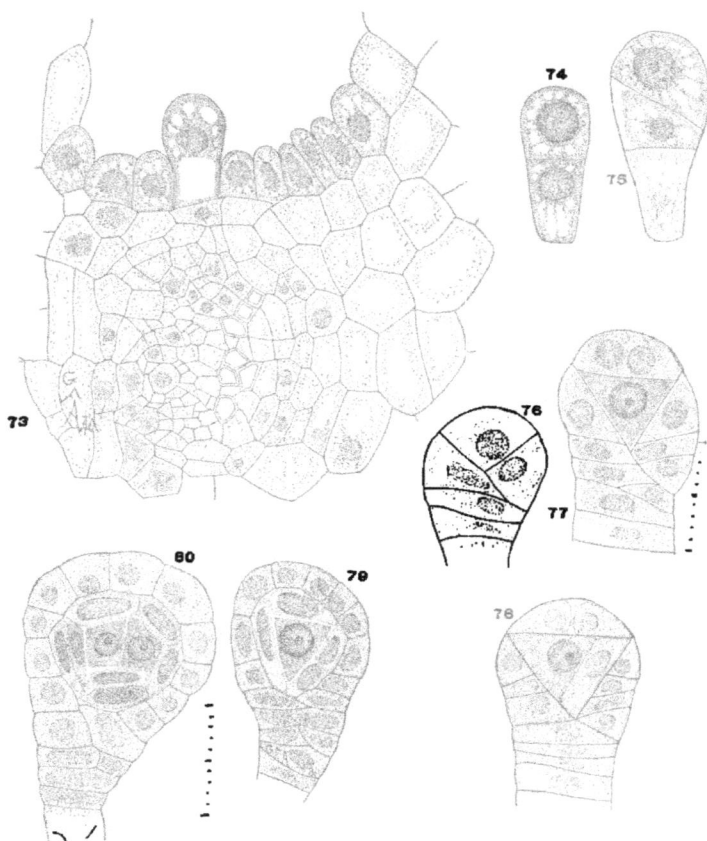

Figs. 73 to 80. Development of the sporangia of Asplenium bulbiferum. Fig. 73, section
through the placental region, showing vascular bundle, epidermal cells rich in chlorophyll, one of them elongating to form a sporangium. Figs. 74, 75, 76, and 77, progressive stages in the farther development. In Fig. 77 the central tetrahedral cell surrounded by the mother cells of the sporangial wall. Figs. 78 and 79, progressive stages
in the development of the tapetal cells, which are cut off as tabular cells from the
four sides of the central cell. Fig. 80, farther division of the tapetal cells. In Fig. 79
the tetrahedral cell is the primordial mother cell of the spores. Magnified 30 times
more than the scale; scale = 1 mm. Figs. 75 and 76 from preparation by O. D.
Humphry; Fig. 77, by H. D. Watson.

where the protoplasm is less granular. Between these radiating granular strands are clear spaces in the protoplasm, comparatively small in the outer end of the cell, but in the base
of the cell very large. Fig. 73, from a section through the
placental region of *Asplenium bulbiferum*, shows in detail
some of the various phenomena described. By a transverse
wall near the surface of the placenta this cell now divides into
two cells. The basal cell develops into the stalk, while the
terminal one is the mother cell of the sporangium, is richly
filled with protoplasm, and contains a large nucleus. The next
line of division in the terminal cell is an oblique one. This is
followed by another oblique wall in the apical cell which joins
the former one at an angle of about sixty degrees. A third
oblique wall now joins these two, forming an end cell shaped
like an inverted, three-sided pyramid. This is followed by a
transverse wall in the tetrahedral cell, which abuts all three of
the others and forms a central tetrahedral cell, very rich in
protoplasm and possessing a large nucleus in comparison with
those of the superficial cells.

By successive oblique walls a tabular cell is cut off from each
side of this central cell, leaving a tetrahedral cell in the centre
which is the primordial mother cell of the spores, or archesporium. The tabular cells surrounding it divide farther, forming one or two rows of cells between the archesporium and the
layer of cells forming the wall of the sporangium. These are
termed the tapetal cells, collectively the tapetum. Meanwhile
the cells of the sporangial wall undergo division by perpendicular walls, thus increasing the superficial extent to accommodate
the increasing size of the contents, but they remain in a single
layer. Over one edge a straight row of cells is differentiated
to form the annulus, while the pedicel develops usually into
three rows of cells.

The tetrahedral archesporium now divides by successive bipartitions into sixteen spore mother cells, with large and
prominent nuclei. Meanwhile the tapetal cells are being

dissolved into a mucilaginous substance in which the spore mother cells float free within the sporangium. The nuclei of the tapetal cells seem to resist the change to a mucilaginous substance much longer than does the cytoplasm, since they may frequently be seen entangled in a disorganized substance collapsed about the spore mother cells.

The steps by which these spore mother cells develop into the spores vary even in the Polypodiaceæ. It will be more convenient to first describe the more common method, not only because it is of more frequent occurrence, but also because it is the method which the writer has observed.

The finely granular content of the nucleus becomes coarser, the granules apparently becoming denser and fewer in number. These increase in size, showing distinct intermediate places of a more homogeneous nature. The granules also coalesce into groups, at first of very irregular, elongated shapes and of various sizes.

The coalescence of these groups continues ; the groups become more and more elongated, and tend also to become arranged nearly parallel, until finally the nucleus has become elongated, elliptical in outline, and the granular part arranged into nearly parallel fibrillæ which converge at the poles of the nuclear spindle. Now, the granular fibrillæ begin to disintegrate in a transverse plane across the centre of the spindle. This disintegration proceeds each way, the disappearing fibrillæ becoming irregular in form, and the entire figure becoming somewhat dumb-bell-shaped, until finally the nucleus is completely divided, and the two smaller nuclei stand off at opposite sides in the cytoplasm of the mother cell, the position being near that of the poles of the former nuclear spindle. No cell wall has appeared separating the cytoplasm.

Each of these nuclei now undergo division in like manner, so that the cytoplasm of each mother cell contains four nuclei. The nuclei now increase in size, granular protoplasm accumulates about them, walls are formed about each quadrant, and

eventually the spores are mature.   Figs. 84 to 90 show various
details of this process.   A few cases which I observed seemed

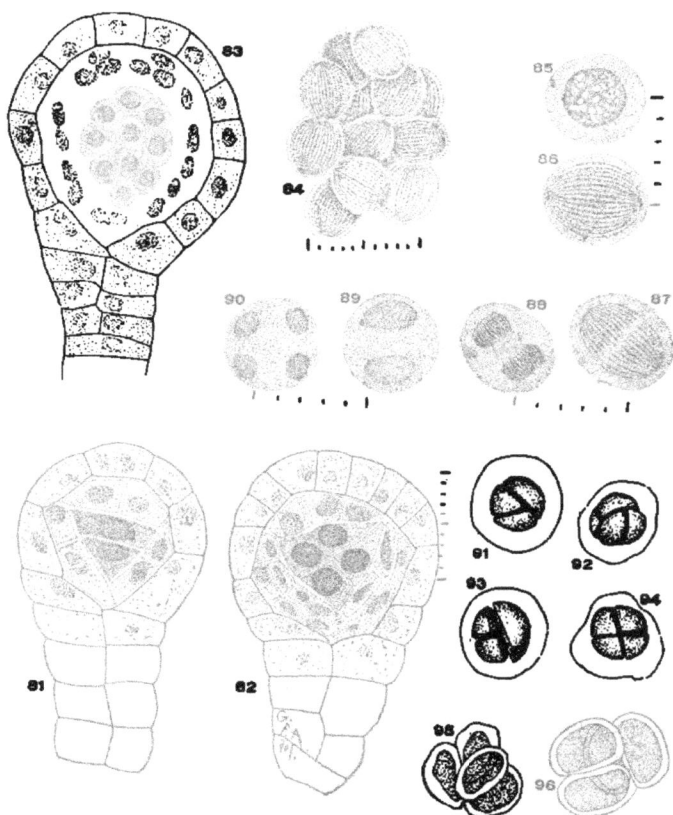

Figs. 81 to 96.   Farther development of the tapetum, its dissolution, and the development of the spores.   Figs. 81 and 82 from Pteris albo-lineata, 91 to 94 from Polypodium
vulgare, the remainder from Asplenium bulbiferum.   See text for details.   Figs. 83 to
90, and 95 and 96, from preparations by E. G. Merritt.   Figs. 85 to 90 magnified 30
times more than the scale; scale = .5 mm.   The remaining figures magnified 30 times
more than the scale; scale = 1 mm.

to show that, in *Polypodium vulgare*, the four nuclei came quite
close together in the centre of the cytoplasm.   As they increase

in size, and the granular protoplasm accumulates about them, the close pressure of the four causes them to form flat surfaces at their planes of junction. The two inner faces of each forming spore meet at the inner angle perpendicularly, forming a rectangle, while the outer face of each forming spore is convex. Two spores in every group were always so arranged that their longitudinal axes were parallel. Sometimes all four were thus arranged. In the majority of cases, each two pairs lay obliquely or perpendicularly across each other. Figs. 91 to 94 show some of the details of these forming spores. The spores are thus arranged radially, and are said to be radial or bilateral.

According to the other plan of development of the spores from the mother cells, each mother cell by successive bipartitions becomes divided into four cells. The protoplasm in each of these now forms a new wall which is the wall of the spore. The original walls forming the divisions of the mother cells dissolve, and set the spores free. These spores are of a rounded or tetrahedral form. This mode of spore formation is said to occur invariably in the Cyatheaceæ, Hymenophyllaceæ, and the Osmundaceæ, while the former occurs for the most part in the other families. In the Schizæaceæ and Osmundaceæ, the sporangia arise from superficial cells on the margin of the leaf before the differentiation of the epidermis.

Structure of the Sporangia. Polypodiaceæ. — In the Polypodiaceæ, the mature sporangium is a somewhat flattened obovoid capsule supported by a long pedicel. It consists of a single layer of cells. The lateral walls are very thin, and present a convexity of surface approaching that of the faces of a biconvex lens. Their margins are separated for about two-thirds the distance by a single row of thicker, nearly cuboid, cells, which form an incomplete ring termed the annulus, prominent upon the back and vertex of the sporangium, its proximal end attached to the pedicel. The front of the sporangium consists of thin cells extending from the distal end of the annulus to the front attachment of the pedicel, and continuous with the lateral

faces. The more central of these frontal cells extend quite far
back, forming a considerable portion of the lateral walls. Two
of these immediately in the centre of the front are so modified
in form as to guide the cleavage of the sporangium at the
moment of dehiscence. These extend farther back over the
lateral aspect of the sporangium than the other frontal cells;
they meet in a straight line, thus offering a direct path for the
cleavage; while the line of their union with the cells directly

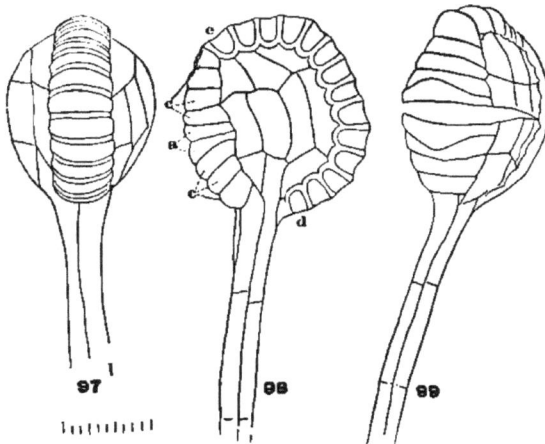

Figs. 97 to 99. Three different views of the sporangium of Aspidium acrostichoides.
a, lip cells; *c*, connectives; *d* to *e*, annulus. Magnified 10 times more than the scale;
scale = 1 mm.

above and below is a sinuous one, the upper lip cell arching
upward at its middle, and the lower one arching downward.
The opening between these two lip cells is termed the stomium,
and a front view of the sporangium presents a semblance to
that of the lips and mouth of some animal.

The thirteen or more cells, arranged in a linear series, and
comprising the annulus, possess a peculiar structure. Their
inner and transverse walls are much thickened, while the lateral

and dorsal walls are very thin and flexible.   The inner walls
form a continuous band with undulations formed by a slight
arching inward of each cell.   The margins of this band are
crenate, the crenations cor-
responding in number to
the number of the cells of
the annulus.   Figs. 97 to
99 show the details.

Cyatheaceæ. — The cla-
vate or obovate sporangia
are short or long stalked.
They are collected in sori
frequently upon a more or
less prominent columnar
placenta at the end of
veins.   Some are naked,
but more frequently they
are surrounded by an open,
circular, cup-shaped indusium, or enclosed in an
ovate   indusium   which
cleaves by a vertical slit
and exposes the sporangia.

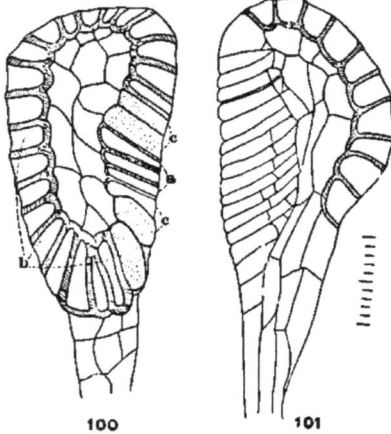

Figs. 100 and 101. Right and left view of sporangium of Cyathea brunonis. *b*, annulus; *a*, frontal series of cells (lip cells) ; *c*, connectives. Magnified 10 times more than the scale; scale = 1 mm.

Goebel says the sporangia possess a complete annulus.   It
is oblique, running vertically around the asymmetric flattened
sporangium.

In *Cibotium chamissoi* H. & B., which I have examined from
the Horace Mann Herbarium of Cornell University (from
Hawaii, distributed in 1867 from the Kew Herbarium), the
annulus is nearly complete, but not entirely so.   It varies in
width and strength, the broader and stouter portions extending
over the base, dorsal, and vertical surfaces.   On the front it is
composed of much smaller cells with thinner lignified walls,
and is interrupted in two places, one at the anterior basal angle,
and another above the middle of the front.   These cells pos-

sess thin walls, are in all respects like those of the lateral walls
of the sporangium, in age sometimes possess brown contents,
and are frequently collapsed. They form a
connective between the ends of the true
annulus and the frontal series of cells. The
line of cleavage proceeds between two cells
of the frontal series. Two lateral views of
the sporangium of *Cibotium chamissoi* are
shown in Figs. 103 and 104.

In *Cyathea brunonis* there is a nearly complete annulus which runs also vertically at a
small, oblique angle around the flattened, cla-
vate, asymmetrical, short-stalked sporangium.
The annulus is interrupted in two places, at
the anterior basal angle, and just above the
middle of the front, these portions forming
the connectives. Sometimes the cells of the
connective possess stouter walls on one side,
but they are very much thinner than those of the true annulus,
and also thinner than those of the short frontal segment

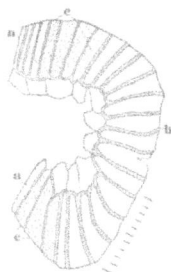

Fig. 102. Sporangium
of Hemitelia spe-
ciosa. *a*, frontal
series of cells; *c*,
connectives; *b*, an-
nulus. Magnified 6
times more than the
scale; scale = 1
mm.

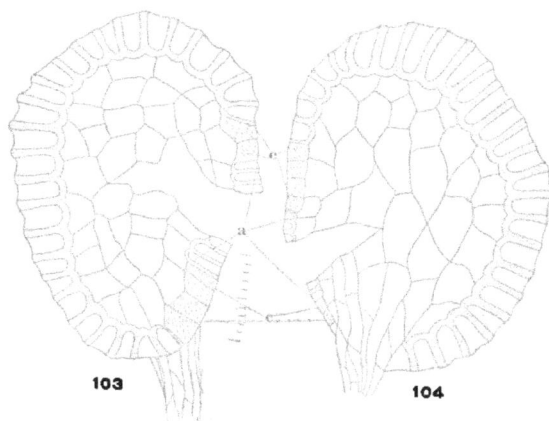

103　　　　　104

Figs. 103 and 104. Right and left view of sporangium of Cibotium chamissoi. *a*, frontal
series of cells; *c*, connectives. Magnified 10 times more than the scale; scale = 1 mm.

between the individuals of which the cleavage takes place. In some cases observed, two different walls in this frontal segment showed the line of cleavage marked out, but it probably would have taken place only in one. The number of cells in this segment varies also as in *Cibotium chamissoi*. The right and

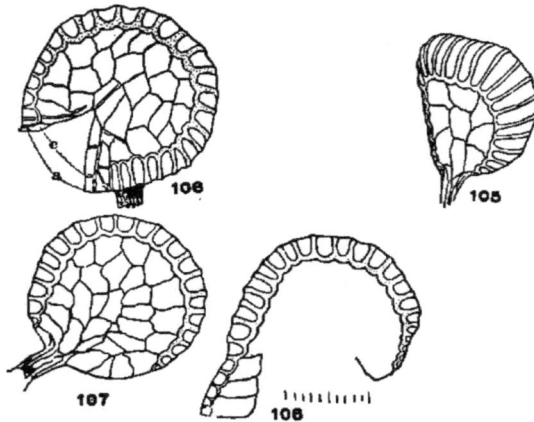

Fig. 105. Sporangium of Hymenophyllum demissum. — Figs. 106 to 108. Sporangia of Hymenophyllum ciliatum. *a*, cleavage cells (lip cells) ; *c*, connectives. Magnified 6 times more than the scale; scale = 1 mm.

left faces of the sporangium are shown in Figs. 100 and 101 ; the annulus is seen to pass close by the side of the base at the point of attachment of the pedicel, which is at one side.

Hymenophyllaceæ. — The sporangia are collected into sori at the ends of fertile leaves or lobes, protected by an indusium in the form of a flap. A more or less elongated columnar or pointed placenta projects as a continuation of the vein of the pinnule bearing the sporangia in short rows. The indusium with the terminal portion of the pinna present the appearance of two slightly broadened laminæ, entirely or only partly enclosing the sorus. The sporangia are sessile, or short-stalked, depressed, asymmetrical, and possess a horizontal, nearly complete annulus. The short pedicel is at one side of the depressed sporangium, the ends of the annulus approaching just above the pedicel. At this point on the upper angle of the sporangium are the elongated, narrow cells with lignified walls which mark the line of cleavage. A few thin, irregular cells with thin walls like those of the other parts of the sporangium form the connectives, so that the annulus is almost complete.

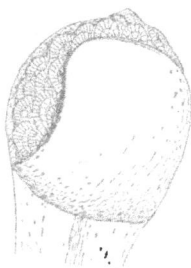

Fig. 109. Sorus of Hymenophyllum demissum.

Gleicheniaceæ. — The sporangia are sessile, naked, three or four in a sorus on the under side of the leaves. They are somewhat depressed, obovate, and asymmetric. The broad annulus, usually said to be complete, extends around the sporangium obliquely or transversely. In *Gleichenia emarginata* the annulus extends obliquely around the sporangium, which usually arises

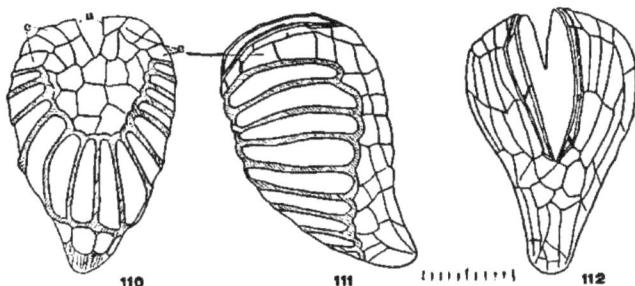

Figs. 110 to 112. Three different views of sporangia of Gleichenia emarginata. *a*, cleavage cells; *c*, connectives. Magnified 6 times more than the scale; scale = 1 mm.

at an oblique angle from the leaf. It presents a broad series of cells with stout, perpendicular walls. It passes underneath the sporangium near the point of attachment with the leaf, and extends up the sides to near the vertex, where it is interrupted on each side by a group of cells forming the connectives. At maturity they sometimes collapse, tend to harden, and are like those of the other parts of the sporangium. They are shorter than those of the annulus, overlap at their ends, lie in several rows, and extend outward to form an attachment to the edges of the series of elongated cells where cleavage takes place. The cleavage cells extend across the vertex and down the front of the sporangium. All the cells of the true annulus arch outward, since the course of the sporangium over which they extend is strongly convex.

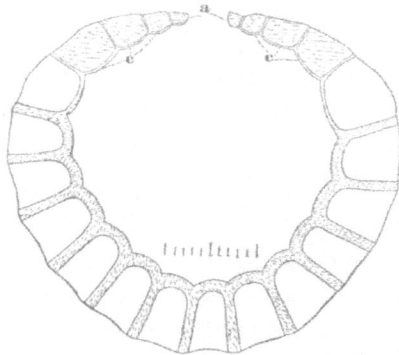

Fig. 113. Section of the annulus of Gleichenia emarginata. *a*, cleavage cells; *c*, connectives. Magnified 10 times more than the scale; scale, = 1 mm.

Schizæaceæ. — The sporangia are oval and sessile. They are naked in *Schizæa* and *Aneimia* and enclosed in a pocket-shaped indusium in *Lygodium*. They are arranged in two rows on the under side of narrow pinnæ. The pinnæ in *Schizæa* have the appearance of a crowded crescent-shaped panicle, while in *Aneimia* the panicle is loose and spreading, and in *Lygodium* the fertile pinnæ are contracted and forked and form a loose panicle. The sporangia are asymmetric, bulging out on one side. The annulus is horizontal, and extends around the conical apex. On the bellied side of the sporangium it consists of narrow and very long cells which extend down to the base in several rows, marking the line of cleavage, which is thus seen to be vertical. The true annulus consists of narrow V-shaped cells, their apices joining the very small plate of thin cells which form the vertex of the sporangium. From this point the V-shaped cells of the true annulus radiate down over the sides of the cone. The wall closing the large part of the V-shaped cells is also thickened, and sometimes the wall of overlapping cells below is thickened for a short distance. Toward the front on either side a few cells of the annulus stand in one or two rows, thus extending farther down the sides of the sporangium, and uniting with the narrow connectives for some distance. A similar structure of the annulus was observed in *Lygodium palmatum* and *Aneimia demissum.* The annulus is thus situated at the opposite end of the sporangium from the

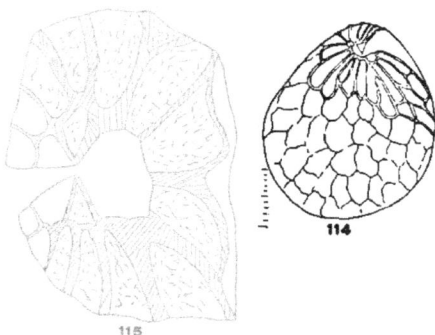

Fig. 114. Sporangium of Schizæa pusilla. Magnified 6 times more than the scale; scale = 1 mm. — Fig. 115. Section of annulus of Schizæa pusilla. The annulus being in the form of a segment of a cone, the transection runs obliquely through the inner wall. Magnified 30 times more than the scale; scale in both figures = 1 mm.

point of attachment of the sporangium to the leaf, and is perpendicular to the vertical line of attachment. The middle part of the annulus is also on the same side of the sporangium as its point of attachment, the front of the sporangium bellying out on the other side. A transection of the annulus of *Schizæa pusilla* is shown in Fig. 115. Since the annulus represents, when viewed from the side, a section of a cone, the inner walls of its cells in transection would appear very broad, since it would be impossible to make, with a straight cut across the annulus, a section

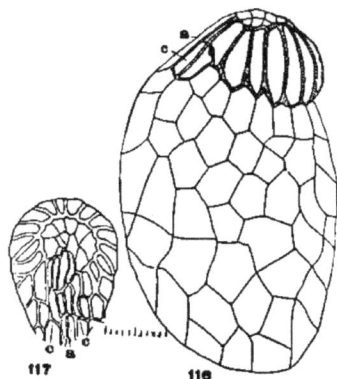

Figs. 116 and 117. Lateral and vertical views of the sporangium of Aneimia phyllitidis. *a*, cleavage cells; *c*, connectives. Magnified 10 times more than the scale; scale = 1 mm.

perpendicular to its inner wall which would at the same time
be transverse to the annulus.   The figure, therefore, shows
an oblique section through the wall.   Fig. 117 is from a section
of the annulus of *Aneimia phyllitidis*.   The section is perpendicular to the inner wall, but not to the entire extent of the
annulus.

Osmundaceæ. — The sporangia are rounded, obovate, short-
stalked, or sessile.   They are asymmetric, possessing a very simple annulus on the dorsal surface of the broad vertex.   The

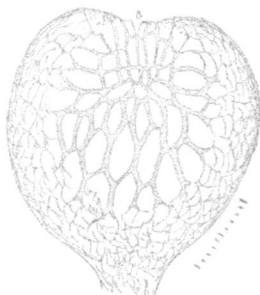

narrow, elongated cells with lignified
walls, which mark the line of cleavage
of the sporangium, extend perpendicular to the annulus, several in parallel
rows, down on the front of the sporangium, which bellies out from the short
pedicel or point of attachment.   The
cells of the annulus form about two
rows transversely across the dorsal
side of the vertex, their longer diam-
eters radiating from the vertical side
transversely to the annulus, and thus
vertically on the sporangium.   The
annulus is connected with the cleavage
cells at the vertex by rather small cells

Fig. 118.   Dorsal view of the
sporangium of *Todea rivu-
laris*, showing annulus and
cleavage cells. Magnified 6
times more than the scale;
scale = 1 mm.

with lignified walls, farther down by cells similar to those of
the lateral walls of the sporangium.   The cells bordering the
annulus below and overlapping its cells have their walls lignified
for a short distance at the point of union with the annulus.

The internal walls of the annulus are very thick and form
a stout plate upon which the bases of the fulcra of the annulus
are fixed.   The fulcra, the longer or transverse walls of the
annulus, are shown in transection of the sporangium (longisection of the annulus) to be quite broad at the base, tapering outward to a rather thin edge at their union with the superficial
wall, appearing in the section narrowly triangular.   The entire

longisection of the annulus presents a crescent figure, the convexity of which is more strongly curved than the concavity, while the intervening perpendicular fulcra become shorter from the middle toward each end.

The transection of the annulus (longisection of the sporangium) presents a somewhat different figure. It is shorter and has the outline of a small segment of a circle, the inner wall being straight. The section being parallel with the long diameter of the cells of the annulus might be several cells deep, when it would present quite a number of fulcra, the oblique perpendicular walls of the annulus. They do not all stand perpendicular to the inner plate, but lean at various angles, reminding one of an arch bridge span. The external wall of the annulus appears to be slightly lignified, but very much thinner than the other walls. The annulus here is also transverse to the line of the short pedicel, its middle part

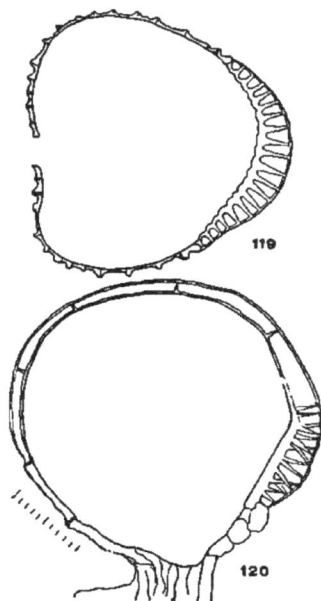

Figs. 119 and 120. Transverse and longitudinal sections through the annulus of the sporangium of Osmunda regalis. Magnified 6 times more than the scale; scale = 1 mm.

being on the same side of the sporangium as in the Gleicheniaceæ and Schizæaceæ. The short annulus is, however, from the shape of the sporangium, situated more nearly across the middle.

VI.

DEHISCENCE OF SPORANGIA AND DISPERSION OF SPORES.

Polypodiaceæ. — In the Polypodiaceæ dehiscence of the sporangium is brought about by the everting of the annulus, which is permitted by the unequal flexibility of its unequally thickened cell walls. While the opening of the stomium between the lip cells is aided by their peculiar form, it seems possible that at maturity the line of union is less firm than between the other cells. The fissure once started proceeds across the lateral walls of the sporangium usually in a straight line, thus splitting in half the cells of the middle row, their frailty favouring this. The drying of the annulus brings about the unequal tension on its cell walls. During this process it slowly straightens, carrying between the distal portion of the lateral walls of the sporangium which remain attached to the free extremity, the greater number of the spores. When straight it continues to evert, and thus usually proceeds until the two ends of the annulus nearly or quite meet, when with a sudden snap it throws the spores violently away and returns nearly to its normal position, with the sporangium nearly closed again. Where the annulus completely everts before a return thrust is made, the drying has proceeded evenly throughout its length. It frequently happens that the annulus dries unevenly; a short segment having dried will snap, while the remaining portion continues to straighten or evert. Several sections of the

68

annulus may thus act, as it were, independently, there being
several distinct snaps or "jerks" in succession.

After this mechanical function of the annulus has been once
performed, and the annulus is dry, it will not take place again
until after the absorption of moisture it dries again, when the

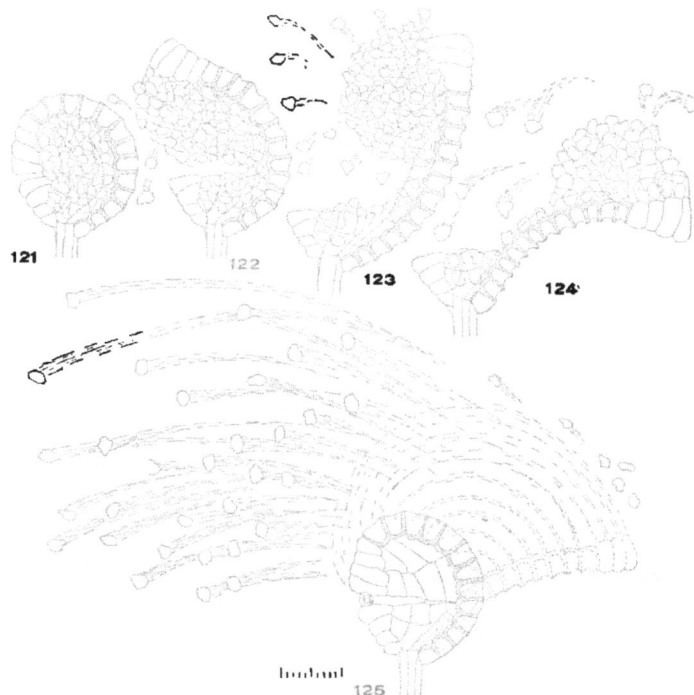

Figs. 121 to 125. Dispersion of the spores from sporangium of Aspidium acrostichoides,
   showing different stages in the eversion and snapping of the annulus. Magnified 10
   times more than scale; scale = 1 mm.

performance is repeated.   This is quite a wonderful and effective provision for the mechanical dispersion of the spores.
With successive periods of wet and dry weather, or humid and
dry air, the annulus repeats this eversion, followed by the vio-

lent snapping, so that the spores which are not at first cast away
will be at a later period.   This process goes on in nature until
the annulus is literally "worn out" by everting, so that the
flexible cell walls become stiffened by age and the accumula-
tion of foreign substances. The annulus will even then respond
to changes in moisture content, but it does not snap suddenly.

The dehiscence of the sporangia and dispersion of the spores
can be accelerated by artificial drying, and beautiful demonstrations of the mechanism be made.   After observing that

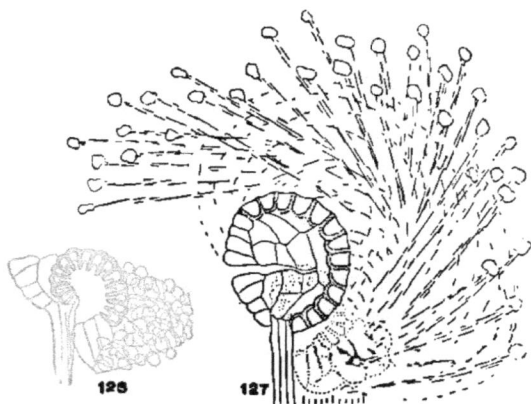

Figs. 126 and 127.   Aspidium acrostichoides.   Fig. 126, complete eversion of the
annulus before the springing occurs. Magnified 10 times more than the scale;
scale = 1 mm.

material dried for several weeks could be moistened and used
for demonstrations of the mechanical function of the annulus,
the writer examined quite a number of species of ferns in the
herbarium of Cornell University, and found that the following
species responded promptly to the test of moisture followed by
drying with artificial heat.   The dry sporangia were scraped
from the sori with a scalpel to a glass slip, moistened with a
drop of water, and dried over an alcohol flame.   Just as the
moisture disappeared from the glass slip, the annuli could be

seen to evert slowly, and then snap into position again with such violence as in many cases to throw them from the field of view.   The following species, collected in the vicinity of Ithaca, N.Y., fourteen to twenty years ago, showed great activity of the annulus : *Asplenium filix-fœmina, angustifolium*, and *thelypteroides*, eighteen years old ; *Asplenium trichomanes* and *ebeneum*, twenty years old ; *Aspidium marginale* and *acrostichoides*, eighteen years ; *Aspidium thelypteris* and *spinulosum*, fourteen years ; *Pellæa atropurpurea*, eighteen years ; *Pteris aquilina*, fifteen years.   The following showed little activity :

Figs. 128 to 131.   Aspidium acrostichoides; different stages in the opening and springing of a free sporangium.   Magnified 10 times more than the scale; scale = 1 mm.

Aspidium goldianum, nineteen years old ; *Aspidium noveboracense*, eighteen years.   In the following, after moistening, many of the annuli would straighten and partially evert, but on drying would slowly return : *Aspidium cristatum*, collected Sept. 9, 1880; several different specimens of *A. cristatum*, collected in 1879.   They appeared as if collected after they were badly weathered, while sporangia of *Aspidium bootii*, collected in September, 1879, would not open, probably never having ripened.   Specimens of *Polypodium vulgare*, matured in 1892 and examined in 1893, after weathering in the meantime, showed slow movements of the annuli, but no sudden return.

Cyatheaceæ. — In the Cyatheaceæ the dehiscence is very similar to that in the Polypodiaceæ. Specimens of *Cyathea brunonis* in the Horace Mann Herbarium of Cornell University, collected in the Malayan Peninsula, and distributed from the Kew Herbarium in 1867, on treatment of the sporangia, showed active annuli, which completely everted and returned with a sudden thrust. The sporangium normally divides across the middle line, guided at one place or another through the frontal series of cells.

Hymenophyllaceæ. — In the specimens of the Hymenophyllaceæ examined very few of the sporangia had dehisced, and in those which had opened, the large majority of spores still remained in the sporangium. Attempts to induce the mechanical action of the annulus were almost complete failures. In many cases the sporangia would partially open, but in only one case was it observed that the annulus sprung, and this case was very feeble. In all the other cases the return was very slow. Nearly all the spores were retained in the sporangium. It is well known that the natural habitat of the Hymenophyllaceæ is in damp situations and humid atmosphere. This may account for the fact that nearly all the sporangia contained their spores, since in a moist or humid atmosphere the sporangia and annulus would have little opportunity to dry out thoroughly. The long time to which the annulus might be subject to moisture might reduce the flexibility of its cell walls so that it would not respond readily to treatment. Bower also calls attention to the fact that very frequently the spores germinate while still within the sporangium. This also goes to show that the spores in the Hymenophyllaceæ are not very effectively dispersed.

The short pedicel being attached near one end of the annulus at one side of the edge of the sporangium, the line of dehiscence passes by the side of the pedicel, guided by the cleavage cells, and divides the sporangium into halves.

Gleicheniaceæ. — In *Gleichenia emarginata* specimens at least thirty years old showed a very active condition of the annulus. Only those cells which above are described as constituting the true annulus showed any of the characteristic changes, or flexibility of the walls, which is the property of the annulus. The opening was different, however, from that in either the Poly-

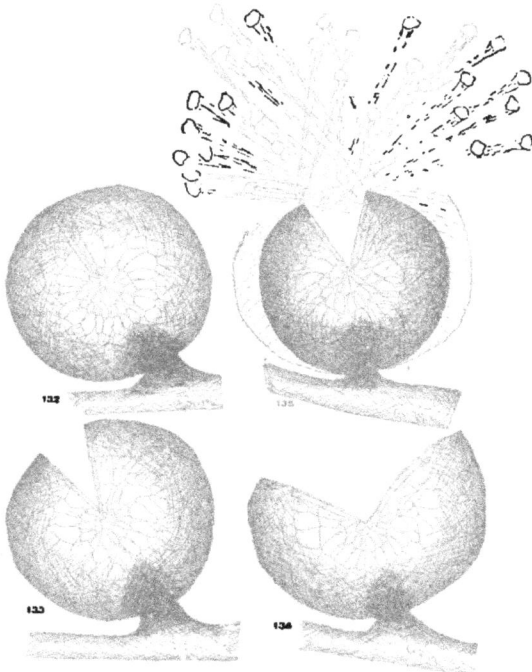

Figs. 132 to 135. Different stages in the dehiscence of the sporangia of Osmunda regalis. Magnified 6 times more than the scale; scale = 1 mm.

podiaceæ or Cyatheaceæ, since the latter possess a vertical or nearly vertical annulus, extending in nearly the same plane as the pedicel. In those two orders the pedicel holds the lower part of the sporangium fixed, unless it becomes separated, and the upper half opens out on the end of the everting annulus.

In *Gleichenia emarginata* the annulus is transverse to the line of attachment, the middle part of the annulus being near this point. The two halves of the sporangium then open out equally on each side, and when the annulus is sprung the two halves move nearly together again. The opening is much the same in the Polypodiaceæ and Cyatheaceæ, if the sporangia are free from the pedicel.

In the Schizæaceæ and Osmundaceæ, the annulus being situated transversely to the line of attachment of the sporangium with the leaf, both halves of the sporangium open out and spring much as in the Gleicheniaceæ, and the spores are thrown in two directions, as shown in Fig. 135.

Comparison. — From this study of the structure and dehiscence of the sporangia, it will be seen that there is a remarkable adaptation of various parts to special functions in the dispersion of the spores. The true annulus in all consists of an incomplete ring. By the true annulus is meant that portion which functions as the spring. It consists of a series of cells possessing stout and firm inner and perpendicular walls, and thin, flexible outer walls enclosing a considerable space which can alternately be occupied by water and air under changing conditions of moisture and dryness, and which under these varying conditions yields to the tension in such a manner that it may be partially or completely everted, the relief from this tension producing a sudden return of the annulus to nearly its former position. The lip cells, or cells of the frontal series, are two in number in the Polypodiaceæ; several rectangular ones in a series in the Cyatheaceæ, and perpendicular to the line of cleavage; a few very narrow and long ones, in a series parallel with the line of cleavage, in the Hymenophyllaceæ, Gleicheniaceæ, Schizæaceæ, and Osmundaceæ. The lip cells approach more nearly the form and structure of those of the true annulus in the Cyatheaceæ, less so in the Polypodiaceæ, and far less so in the four other orders. In the Polypodiaceæ, Cyatheaceæ, Hymenophyllaceæ, and Gleicheniaceæ there are two series of cells similar to those of the lateral faces of the sporangium, which serve as connectives between the ends of the true annulus and the lip cells. These are smaller and fewer in number in the Hymenophyllaceæ, and greater in extent in the Polypodiaceæ. In the Osmundaceæ the connective consists of a very few small cells with lignified walls at the vertex of the sporangium, bordered by larger cells similar to those of the lateral walls.

The function of the lip cells is to start the line of cleavage for the dehiscence so that the sporangium shall open by two nearly equal parts. The function of the connective is not only to serve as the pull on the lip cells, and to form an attachment to the lateral walls of the sporangium, but also to hold the lateral walls of the sporangium from going to pieces, in order that this part of the sporangium may retain so far as possible its normal form and position, and hold the spores in place until the annulus springs, when they are violently hurled away.

A complete annulus in these orders would defeat such an effective dispersion of the spores, since the entire length of the line of cells around the sporangium would then evert, and in doing so would tear the lateral walls apart, or completely evert them, and the spores would fall away before the annulus would spring. In the Schizæaceæ, it is also doubtful if the entire series of cells functions as the spring; but if it should, being situated on a small tapering apex of the oval sporangium, it could not completely evert the walls of the sporangium, and so would not operate against the effective dispersion of the spores.

Mechanism of the Annulus. — Prantl, in 1879, described the mechanical function of the annulus, having observed it in the Polypodiaceæ, Cyatheaceæ, Hymenophyllaceæ, Gleicheniaceæ, and Osmunda. In the layers of protoplasm lining the internal surface of the cells of the annulus, he states there is some substance which possesses a strong avidity for water. On placing dehiscent sporangia in water, he observed that the strong endosmotic pressure resulting from the absorption of water in the cells caused the air in their cavities to be quickly absorbed. Upon drying, or by treating with glycerine, alcohol, strong solutions of potash or chloriodide of zinc, or any substance which would withdraw the water, the cell walls not being permeable to air, the pressure from the outside caused the thin dorsal and lateral membranes to infold, and in doing so pulled on the radial walls of the annulus. These acting as fulcra, their outer ends were pulled near together, causing the annulus to be looped in the opposite direction. Finally, when nearly all the water was abstracted, the air held in solution is suddenly set free. The outside pressure thus being overcome, the annulus suddenly returns nearly to its former position. Since plasmolysis does not occur, the substance in the cells which possesses a strong avidity for water does not escape, and the process can be repeated successively.

Where the process goes on evenly in all the cells, the annulus is completely everted before it springs; but when it goes on unequally in different cells, several sections of the annulus may evert and spring independently.

Schinz, in 1883, made the statement that the straightening of the annulus and dehiscence of the sporangium were due to the unequal contraction in drying of the unequally thickened cell walls of the annulus.

In 1885 Leclerc du Sablon gave an explanation similar to that of Prantl. In the same year Schrodt said the opening of the sporangium was due to atmospheric pressure on the thin dorsal and lateral membranes during the abstraction of water; that the sudden return was caused by the sudden entrance of air through the membrane, which moist was not permeable by air of the pressure of one atmosphere. According to his theory, the membrane, reaching a certain degree of dryness, becomes permeable to air of the pressure of one atmosphere. The air which then enters is rarefied, and the annulus does not entirely close. Prantl showed that this theory was untenable; for example, when glycerine was used to withdraw the water from the cells, there was no air surrounding the annulus which could suddenly pass into the cells.

VII.

SUBSTITUTIONARY GROWTHS.

THE normal life cycle, as illustrated by the succession of the gametophyte and sporophyte phases of the fern plant, is frequently interrupted by various substitutionary growths. While the perennial character of nearly all ferns renders it a very common thing for the individual sporophyte to exist for years independent of the gametophyte phase and produce successively crops of spores, this does not necessarily interfere with the normal sequence of the two phases. For, omitting exceptional cases to be described hereafter, the spores produced each year from these perennial individuals are capable, under normal conditions, of intro-

Fig. 136. Sporophytic bud from leaf of Asplenium bulbiferum.

ducing the normal life cycle again, and originating new plants. We might then consider the normal life cycle to be the production of new plants by a succession of these two phases.

It is now known, however, that new plants in many cases may be introduced either by substitutionary growths or by the

77

direct omission or arrest of a part or the whole of one of the
two normal phases.

Sporophytic Budding. — What is called sporophytic budding
has long been known to occur in various species of *Asplenium,
Cystopteris*, etc., where bulbs are formed which are indeed young
plants, and are capable of developing into perfect plants.   In
this case superficial tissues of the sporophyte, usually near a
vascular bundle on the leaves, grow out directly into a new
sporophyte.   Various species of *Asplenium* grown in conservatories afford beautiful examples of this phenomenon.   *Asple-*

Figs. 137 and 138.   Asplenium bulbiferum.   Fig. 137, section of normal pinnule, showing
epidermis, mesophyll cells, and part of vascular bundle.   Fig. 138, section through
pinnule at very young sporophytic bud, showing mesophyll cells, small and rich in
protoplasmic content, and the apical cell of the bud.   Magnified 30 times more than
the scale; scale = 1 mm.

nium bulbiferum, for example, presents frequently numerous
young plants in various stages of growth, and of various sizes.
A single leaf may bear more than a dozen, and frequently
several may be found upon one pinna, one each on several of
the pinnules near the extremity of the pinna.   By examining a
growing leaf, it will be observed that near the tip some of the
very young pinnules are abruptly curved, forming a greater or
less angle near their attachment to the pinna.   If this angle of
such a young pinna be examined with a pocket lens, the surface

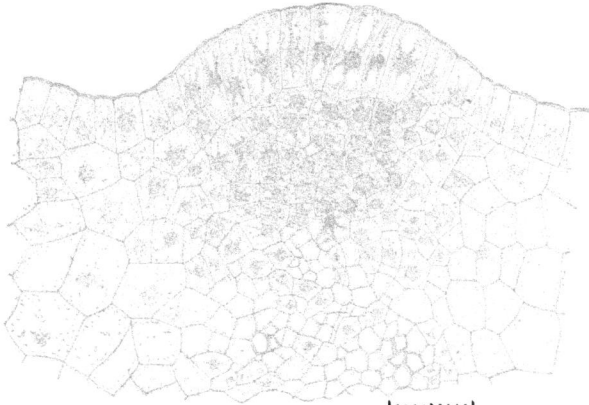

139

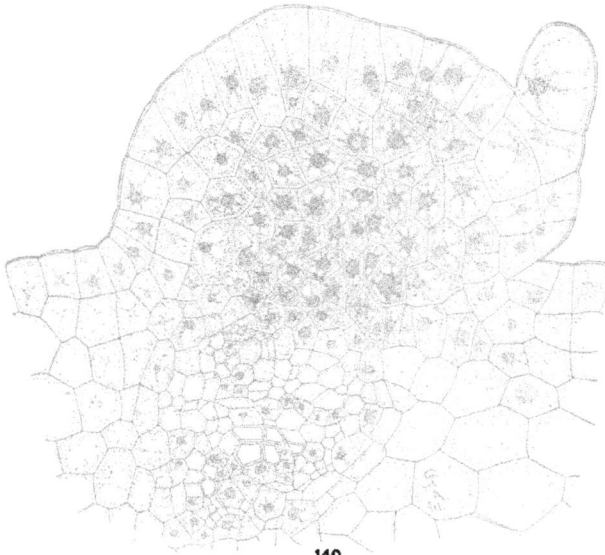

140

Figs. 139 and 140. Sections through young sporophytic buds of Asplenium bulbiferum. Magnified 30 times more than the scale; scale = 1 mm.

of the vein in the angle will usually appear somewhat broader,
and a greater or less convexity of the surface is present, according to the age of the growth.   Sectioning such a pinnule shows,
at this point, a tissue on that side of the vein composed of
mesophyll and epidermal cells, very rich in protoplasm and with
somewhat smaller mesophyll cells than those of the opposite
side of the section, or in other parts of the pinnule.   The

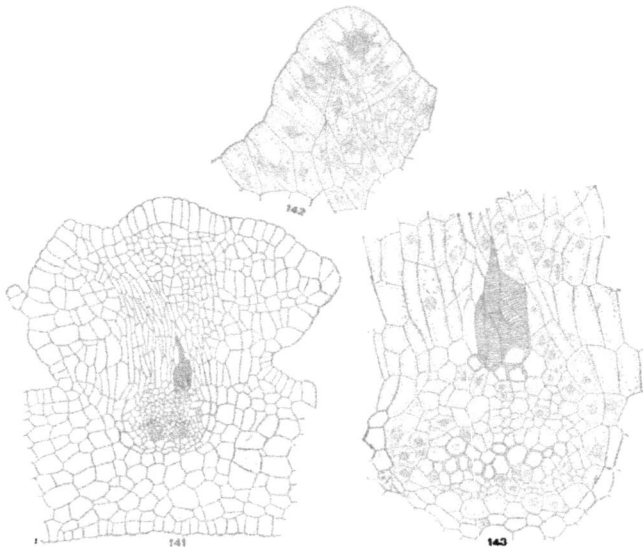

Fig. 141.   Section of young bulb of Asplenium bulbiferum, magnified 6 times more than
the scale. — Fig. 142.   Apex of leaf with apical cell from section adjacent to 141. —
Fig. 143.   Detail of vascular bundle from 141, showing the origin of the tracheides
of the bud. Figs. 142 and 143 magnified 30 times more than the scale; scale = 1 mm.

smaller size of the mesophyll cells, and their rich content of
protoplasm, indicates at once a very different function from the
ordinary mesophyll cells.   The tissue has become a growing
point, or bud, in which are to be found the primary elements of
stem, and later of leaf also.   This changed function of the cells
in this area of the pinnules, carrying with it, according to the

law of correlation, change in the form and size of the cells, accounts for the angle in the pinnule at the point of origin of the bud. The other mesophyll cells have grown to their accustomed size, thus producing unequal tension on the two opposite sides of the pinnule.

Sections at a very early age show what appears to be an apical cell, perhaps of the stem (see Fig. 138), though it is not always well apparent. It would be incorrect, however, to say that the bud is derived from a single epidermal cell, or indeed from the epidermis. The mesophyll cells between the epidermis and the vascular bundle partake also in the growth and cell divisions. Later its cells are the first to elongate, lose their rich protoplasm, and take on the character of scalariform tracheides, which are in immediate communication with those of the bundle of the pinnule. It may be that the bud is derived from a single superficial cell of the young leaf before the differentiation of the epidermis.

As the bud grows, numerous elongated brown scales are developed, which cover or protect it, and later young leaves arise from different places, unrolling in circinate fashion, the earliest ones soon expanding, and the form of the young fern plant is then apparent. These buds on *Asplenium bulbiferum* usually arise from the upper surface of the pinnule, near its base. Occasionally they arise from the under side. Frequently a sorus occurs along the side of the vein by a bulb containing well-developed sporangia and spores, so that no sporal arrest occurs in connection with the sporophytic budding in this fern. Other species of *Asplenium* produce similar sporophytic buds, which have been studied by Heinricher, as *A. celtidifolium, bellangerii, viviparum, decussatum,* etc.

In *Cystopteris bulbifera,* a very common fern in cool, damp rocky glens, the bud while still attached to the leaf does not resemble a young plant at all. It presents more the appearance of a fleshy bulb. Large specimens present an oval form, the larger part of which is composed of two opposite thick rudi-

mentary leaves, elliptical in outline, convex on the outer face
and plane on the inner.   Their outer ends are separated by the
younger rudimentary leaves, two of which can usually be seen
in well-matured buds, and the stem axis.   The basal half of the
bud appears as a consolidated mass of tissue.   All these rudimentary leaves are different in size, since they arise successively.   The buds easily break away and fall to the ground,
where they grow into new plants by an elongation of the stem

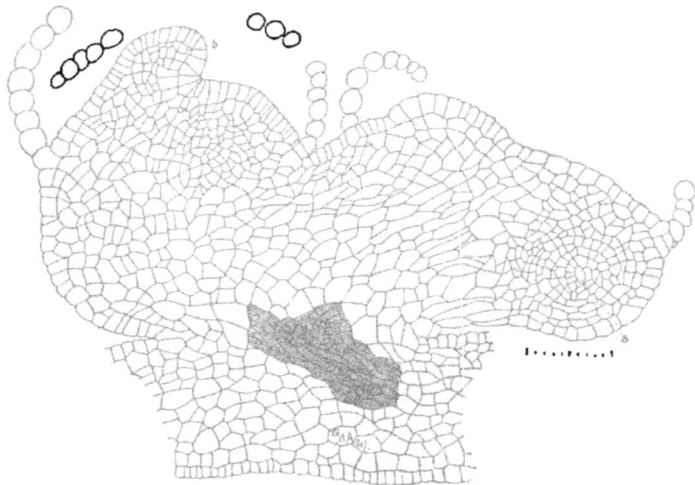

Fig. 144.   Section of bulb of Asplenium bulbiferum.   a, transection of leaf; b, longisection of leaf.   Magnified 6 times more than the scale; scale = 1 mm.

axis and the production of roots and new leaves.   The first
leaves so produced are usually rudimentary, and their stipes
possess very large bases.   The tissue of these rudimentary
leaves in the bud is chiefly composed of large parenchymatous
cells with intercellular spaces and packed with starch grains,
which form reserve material to supply the young plant with
food until it shall gain a foothold in the soil or in the crevices
of rocks.

Longitudinal sections of even quite young bulbs show the vascular bundle of the stem in connection with that of the rachis of the leaf, while leaf bundles branch off from this into the rudimentary leaves. Fig. 147 is from a section of a bud only about one-third grown, showing two rudimentary leaves, the growing end of the stem, and a young root which has not yet emerged from the tissue of the stem. The buds are situated on the under side of the leaves in the axils of the pinnæ.

Bulbs in the axils of the leaves of *Asplenium (Athyrium) filix-fœmina*, var. *clarissima*, and in connection with the sorus in var. *plumosum*, are reported by Drury. Bower records sporophytic buds on the leaves of *Aspidium erythrosorum*, var. *proliferum*, and *Trichomanes pyxidiferum*. They also occur on the base of the stipe in *Pteris aquilina;* in *Aspidium filix-mas,*

Figs. 145 and 146. Cystopteris bulbifera. Fig. 145, bulb in the axil of the leaf; 146, young plant developing from a bulb.

farther up on the lateral edge of the stipe. *Blechnum hastatum* and *Onoclea struthiopteris* produce stolons or adventitious buds. Other cases are such as occur when a leaf is "layered," and which habitually takes place in the walking fern, *Camptosorus rhizophyllus.*

Gametophytic Budding. — This is the development of new prothallia as direct outgrowths from other prothallia. It occurs, as we have seen above, according to Goebel, in *Gymnogramme leptophylla.* According to Rauwenhoff it occurs in *Gleichenia,* and, as the writer has observed, in *Adiantum cuneatum.* Bower

first termed this "oophytic budding," and he cites a case in a fern prothallium observed by Cramer.

There are other departures from the normal life cycle of ferns which consist in the partial or complete arrest of the usual reproductive bodies, and the appearance in their stead of the complementary phase of the life cycle. Such phenomena are known as apogamy and apospory.

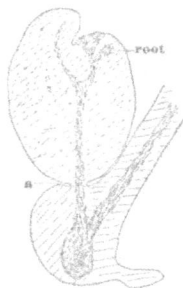

Fig. 147. Diagrammatic section of a bulb of Cystopteris bulbifera made up from several longitudinal serial sections. *a*, point of attachment of the bulb with the rachis of the leaf.

Apogamy. — This was first described by Farlow as occurring in *Pteris cretica*, and is the development of the sporophyte from the gametophyte without the intervention of the sexual organs. The meristematic tissue, instead of developing archegonia, showed scalariform tracheides, and later a development of young fern plants directly from the prothalline tissue. De Bary recorded it in *Aspidium filix mas*, var. *cristatum*, *A. falcatum*, and *Todea africana*. According to these investigators, prothallia of all these species bear antheridia. Archegonia are entirely absent in such prothallia of *Aspidium filix-mas*, var. *cristatum ;* never reach maturity in *Pteris cretica ;* while they may reach maturity in *Aspidium falcatum* and *Todea africana*. Bower has described cases of apogamy in *Trichomanes alatum*, and Berggren on the prothallia of *Notochlæna distans*.

Apospory. — Apospory is the development of the gametophyte from the sporophyte without the intervention of the spores, and occurs in several different ways. This was first reported by Drury in the case of *Asplenium (Athyrium) filix-fœmina*, var. *clarissima*, Jones, and *Polystichum angulare*, var. *pulcherrimum*, Padley, and was later carefully studied by Bower. In *A. f.-f.*, var. *clarissima*, no mature sporangia or spores are developed. Accompanying this complete sporal arrest, the sporangia have taken on the function of producing prothallia directly. This occurs in two ways. Prothallia may be developed from the cells of the wall of the sporangium or from the stalk. In no case does the archesporium partake in this aposporus growth. In certain arrested sporangia some of the cells show a rich content of protoplasm, and later, by growth and division of these cells, the prothallia are developed. The forms of these growths are various, and many of them depart widely from the normal form of prothallia, but their gametophytic nature is assured by the production of antheridia, and in some cases by archegonia.

In *Polystichum angulare*, var. *pulcherrimum*, prothallia are produced from four different places on the sporophyte: from the apex of pinnules not connected with a vein; from the surface of pinnules near a vein; from arrested sporangia; and from the base of the sorus.

Fig. 148. Young fern plant, an apogamous growth from the prothallium of Pteris cretica.

Farlow has described apospory in *Pteris aquilina*. The prothalline growths arise from arrested sporangia on fertile pinnules which are much dwarfed and curled, and on which sporangia much in advance of those on normal pinnæ occurred.

Bower has farther given a very interesting account of apospory in two species of *Trichomanes*. In *T. pyxidiferum* prothallia are produced from the leaf with partial sporal arrest. In *T. alatum* the prothallium is either protonemal-like, or an expanded plate of cells. Protonemal threads may arise from cells of the apex of the leaf, from its margin, or from the surface of the leaf in connection with a vein, or from the sporangia. Later flattened expansions may be produced on these protonemal threads. In other cases, the prothallia as flattened expansions may arise directly from the leaf without the intervention of the threads. Frequently these bear at their apex gemmæ, which are several-celled bodies ovoid or elliptical in form and supported by sterigma. They have been seen to germinate.

OPHIOGLOSSEÆ.

THE prothallia of the Ophioglosseæ are not well known. In the species examined a tuberous body has been found, destitute of chlorophyll, and bearing both antheridia and archegonia. Goebel suggests the probability that a green prothallium is at first developed, much as in the case of *Gymnogramme leptophylla* in the Polypodiaceæ.

In *Botrychium* the antheridia are sunk in cavities in the tissue, while in *Ophioglossum* they only project slightly above the surface. In *Ophioglossum*, according to Goebel, the mother cells of the spermatozoids are developed from one or two cells of the inner tissue of the antheridium. The archegonium is similar in development to that of the homosporous leptosporangiate Filicinæ. The venter is sunk in the tissue of the prothallium, and the short neck projects above.

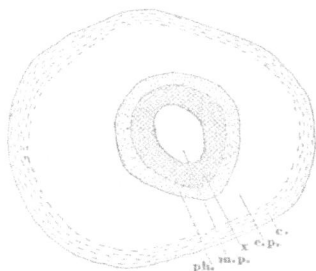

Fig. 149. Diagrammatic transection of the stem of Botrychium virginianum. *x*, xylem; *ph.*, phloem; *m.p.*, medullary parenchyma; *c.p.*, cortical parenchym; *c*, cortex.

Sporophyte. Stem. — The stem of the Ophioglosseæ is very short, sometimes with a bulbous base, and fleshy. It is usually erect and ascending, but in *Helminthostachys* the underground stem is creeping. It rarely branches. The fundamental tissue consists of large, usually cylindrical cells, with prominent intercellular spaces and richly filled with starch. It is separated by the annular bundle into cortical and medullary portions. In age the cortical tissue sometimes develops a deep layer of cork cells. There is no sclerenchyma.

The annular bundle consists of a hollow cylindrical network, with foliar gaps associated with the leaf traces. Sometimes

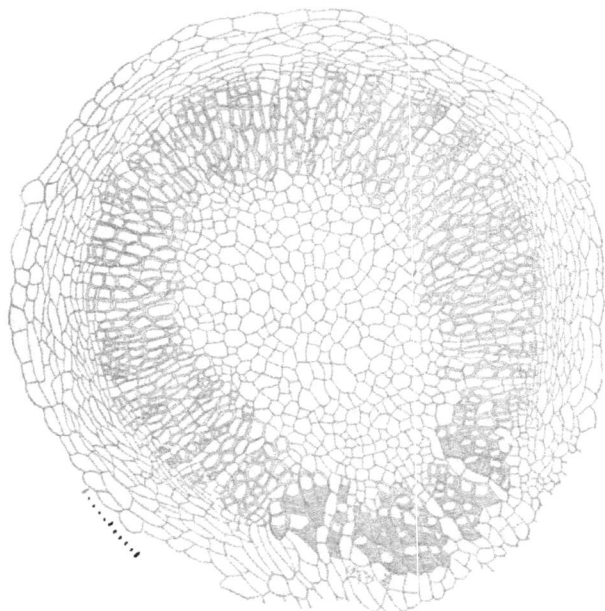

Fig. 150. Section of the annular bundle of Botrychium virginianum, showing the radial bands of parenchyma connecting the medullary parenchyma with the phloem. Magnified 6 times more than the scale; scale = 1 mm.

the tissue of the foliar gaps changes to scalariform tracheides, and a continuous hollow cylinder of xylem results. The xylem of the bundle is on the axial side, the phloem on the peripheral side, and the bundle is thus collateral. In *Botrychium rutæfolium*, according to De Bary, parenchyma is present in the annular bundle in the form of radial bands which resemble medullary rays, while in several specimens of *B. lunaria* which he examined it was not found.

In *Botrychium virginianum*, which I have examined, parenchyma is present in the annular bundle. It appears as radiating bands of one or two rows of cells connecting the medullary with the cortical parenchyma. The radial diameter of these cells is two to three times their tangential diameter. In radial longisection, the band is seen to consist of four to eight rows of cells, or more. The cells are rectangular in outline. Their narrow tangential diameter shows that they have been flattened by the pressure of the xylem.

A rudimentary bundle sheath is present in the stem of *Botrychium*. It can be distinguished by the transverse folding of the middle of the longitudinal radial side walls of its cells.

Leaf. — In the leaf stalk of *Ophioglossum*, the bundles are collateral, the xylem consisting of narrow reticulate tracheides with no intercellular spaces. In *Botrychium*, the bundles of the leaf stalk are concentric, the xylem consisting of scalariform or reticulate tracheides. In *Botrychium* there are four bundles, while in *Ophio-*

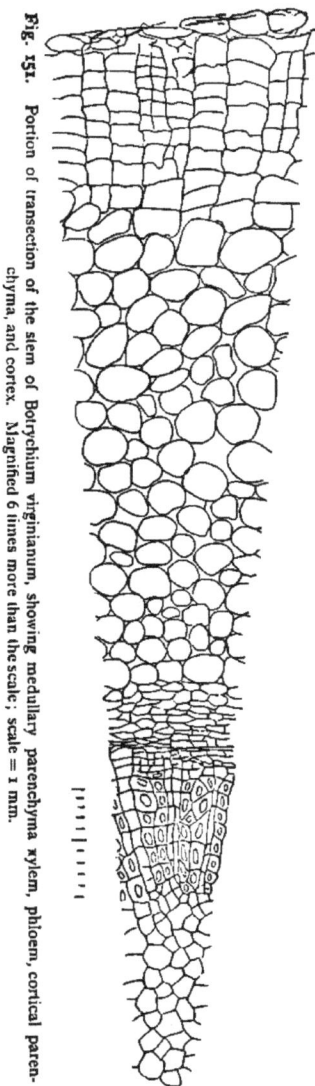

Fig. 151. Portion of transection of the stem of Botrychium virginianum, showing medullary parenchyma xylem, phloem, cortical parenchyma, and cortex. Magnified 6 times more than the scale; scale = 1 mm.

glossum there may be eight.   They are arranged in a circle separated by parenchyma.

The leaves require several years for development ; for *Botry-chium lunaria*, it is said, over four years.   In some species only one leaf is developed each year, while in any case the number is few.   Usually the leaf, in its embryonic development, divides into a fertile and sterile portion, but some species of one section of *Ophio-glossum* possess both fertile and sterile leaves. The fertile portion of the leaf may be simple and entire, as in *Ophio-glossum ;*   simple,   or once, twice, or thrice pinnate or ternate in *Botrychium.* There is a long petiole with sheathing margins at the base, which, equitant fashion, ride upon and enclose the end of the stem and younger leaves.   The leaves are not circinate in vernation, but when young may be straight, partly inclined, or wholly inclined, according to the species.   The epidermis of both surfaces possesses stomates, and the mesophyll is pro-

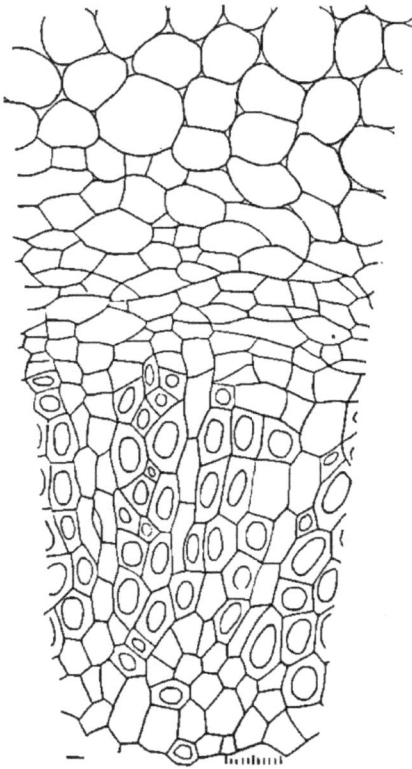

Fig. 152.   Portion of transection of bundle in the stem of Botrychium virginianum : xylem portion, consisting of tracheides, separated by radial bands of parenchyma ; phloem, bundle sheath, and a portion of the cortical parenchyma.   Magnified 30 times more than the scale ; scale = 1 mm.

vided with intercellular spaces.   The bundles of the leaf are weak, are said to branch dichotomously in *Botrychium* and *Helminthostachys*, while they anastomose in *Ophioglossum*.

In a number of species the fertile branch is destitute of chlorophyll.   In a specimen of *Botrychium lunarioides* from Concord, Mass., two divisions of the usually sterile lamina were changed to fertile ones.

Sporangia. — The sporangia are an entire year in completing their development.   They originate in what is called the sporog-

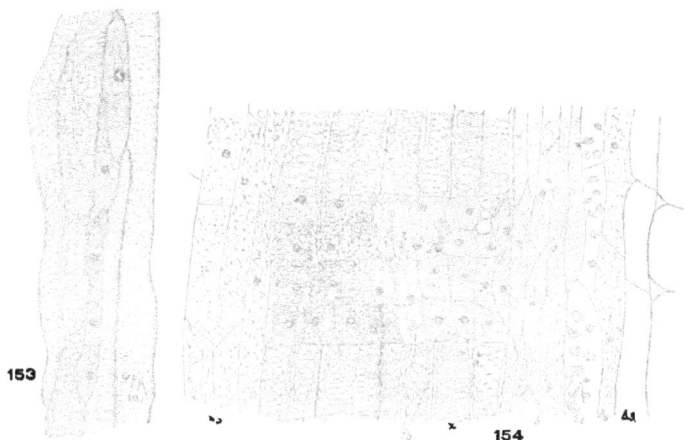

Figs. 153 and 154. Botrychium virginianum. Fig. 153, tangential section of portion of xylem showing a transection of one of the radial bands of parenchyma; 154, radial section of bundle. *b.s.*, bundle sheath; *p.s.*, phloem sheath; *s.t.*, sieve tubes; *x*, xylem; *m.p.*, medullary parenchyma; *r.b.*, radial band of parenchyma. Magnified 30 times more than the scale; scale = 1 mm.

enous tissue of the fertile part of the leaf.   The wall is developed from the epidermis, while a terminal cell of an axial row beneath forms the archesporium, according to Goebel; while Holzman thinks a large superficial cell, in *Botrychium*, present in the early development, may be considered the mother cell of the sporangium, though it does not project above the surface before division.   A tapetum surrounding the spore mother cells is developed from tabular cells cut off from the archesporium. Numerous spores are developed in a single sporangium. The sporangia are modified pinnules of the fertile part of the leaf. In some species, where the leaf is more simple, the sporangia appear on the inner side of the fertile leaf. In *Ophioglossum* the sporangia closely cohere, and the form of the fertile part of

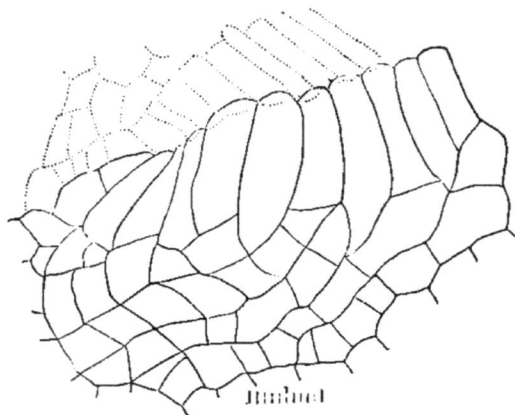

Fig. 155. Longitudinal section of the end of the stem of Botrychium virginianum, showing apical cell. The dotted lines represent the borders of cells of a leaf closely crowding upon the end of the stem. Magnified 30 times more than the scale; scale = 1 mm.

the leaf is a two-ranked spike. In *Botrychium* they are more separated, and once to thrice pinnate.

Roots. — The roots are quite fleshy, usually somewhat brittle, and when dry, spongy, so that they swell to nearly their normal size when soaked in water. They arise one at the base of each leaf trace in the stem. In *Ophioglossum*, the roots do not branch, while in *Botrychium*, a few lateral roots arise in a monopodial fashion. In *Ophioglossum*, many of the roots produce adventitious buds which develop into new plants. The epidermis of the roots consists of cells with hard brown walls. The various tissues are differentiated from a tetrahedral apical

cell, the cells of the root cap, in some species, soon losing the
stratified character so persistent in the roots of ferns.   In

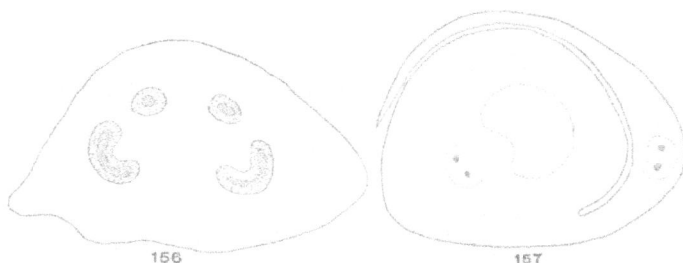

Fig. 156.   Diagrammatic section of the stipe of Botrychium virginianum, showing the
    arrangement of the bundles. — Fig. 157.   Diagrammatic section of the end of the
    stem and equitant bases of the leaves, with leaf traces and open portion of the stem
    bundle (B. virginianum).

Botrychium virginianum the stratification is quite well marked.
Very few root hairs are developed.   The bundle sheath is feebly
developed ; it can be differentiated by the folding of the middle
of the longitudinal radial side walls of its cells.

In *Ophioglossum* the small bundle is collateral.   The xylem,
the cells of which are similar in transection to those of the
stem, in the form of a crescent occupies one side, its convexity
separated by a layer of one or two
rows of delicate cells from the bundle sheath.   The phloem, in several
layers, borders directly on the bundle
sheath on the opposite side of the
bundle.   According to Van Tieghem,
two of the outermost cells of the
xylem sometimes border directly on
the bundle sheath.

In *Botrychium* the bundle is concentric, or perhaps, more correctly
speaking, radial.   In primary roots it
usually consists of three, sometimes

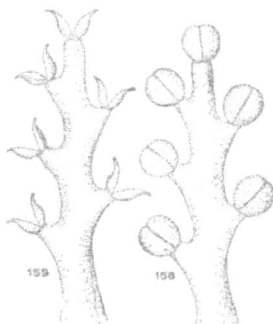

Figs. 158 and 159.   Mature and dehiscent sporangia of Botrychi-
um virginianum.

four, radial narrowly oval groups of xylem alternating with as many groups of phloem. The triarch bundle is the more common, the tetrarch bundle occurring sometimes in stout roots. I have observed the triarch in *B. matricariæfolium* and *B. virginianum,* and the tetrarch in *B. virginianum* and *B. ternatum.* In the lateral roots the bundle is usually diametrically diarch; occasionally it is triarch in *B. virginianum.*

The structure of the elements of the xylem is the same as that in the stem, except that no rays corresponding to those in the stem bundle of *B. virginianum* and *B. rutæfolium* are present. The xylem begins by the appearance of a few small tracheides at two, three, or four points, a few cell layers inside the bundle sheath, according as the bundle is to be di-, tri-, or tetrarch. The transformation of the cells to xylem then proceeds inward from these points toward the centre, the number of rows of xylem cells as well as their size usually increasing on the axial side. In old roots the intermediate parenchyma cells in the centre sometimes become completely obliterated by their transformation to xylem. In one old root of *B. ternatum* of the tetrarch type, the four groups of xylem appeared to have coalesced. Only here and there was a single cell compressed between the xylem, and indicating the earlier four radiating lines of separation. In this case the xylem presented four concave, external faces, the concavities occupied by the phloem.

The development of the phloem proceeds from similarly situated, alternating points with that of the xylem. From each point the transformation of the procambium then proceeds in three directions as seen in cross-section, one toward the centre, and two tangentially. According to the age of the root, more or less of the tissue intervening between the lateral faces of the xylem is thus transformed to phloem. It consists of thick-walled cells which, untreated, possess a bright gleam viewed by transmitted light, and by treatment take a deep stain. Sieve tubes occur.

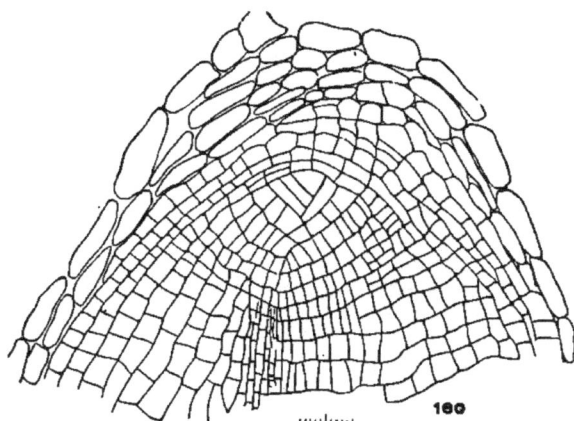

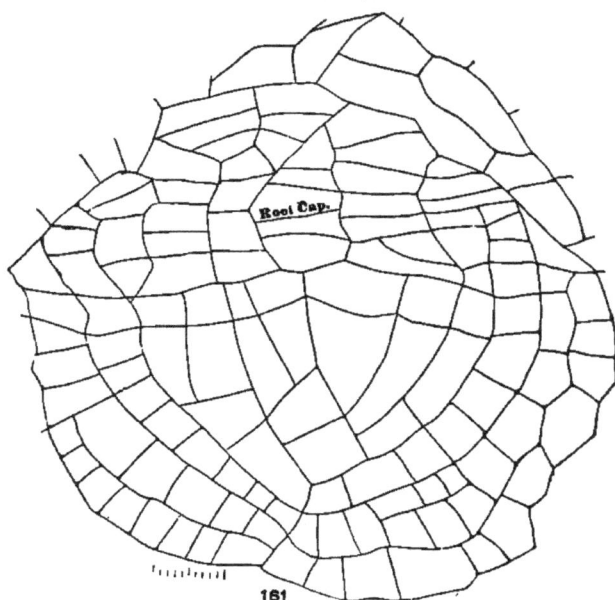

Figs. 160 and 161. Longitudinal section of root tips of Botrychium virginianum, with apical cell and root cap. Fig. 160, from mature root, magnified 20 times more than the scale; 161, from young root still within the stem, magnified 30 times more than the scale. Scale = 1 mm.

The xylem is separated from the bundle sheath by one or two layers of large, thin-walled cells, and even in old roots there appears at least one layer of such cells between the phloem and bundle sheath.

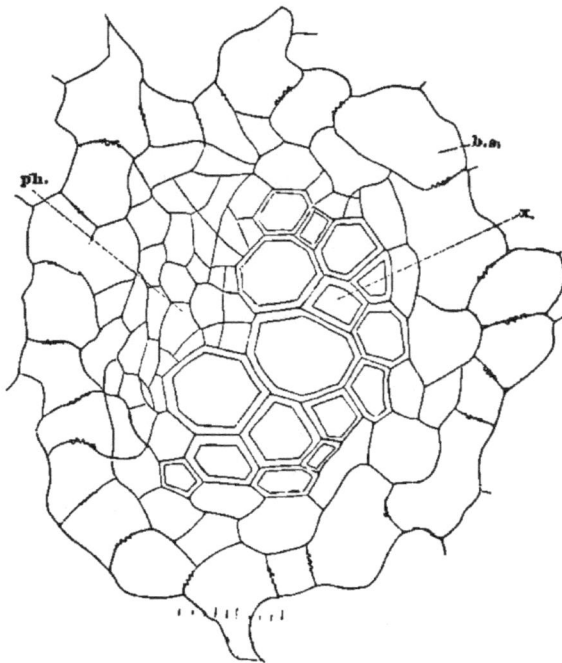

Fig. 162. Transection of the bundle of the root of Ophioglossum vulgatum. *x*, xylem; *ph.*, phloem; *b.s.*, bundle sheath. Magnified 6 times more than the scale; scale = 1 mm.

The fundamental tissue is made up of deep layers of cortical parenchyma, richly filled with starch, and possessing large intercellular spaces.

As the writer has shown, symbiosis occurs in connection with the roots of the Ophioglosseæ. A filamentous fungus occupies areas of varying extent, at a definite distance from the surface of the root. From this localized centre of metabolic activity threads of the fungus extend to the outside of the root. The presence of this microsymbiont probably bears a close relation to the almost complete absence of root hairs in the order. The

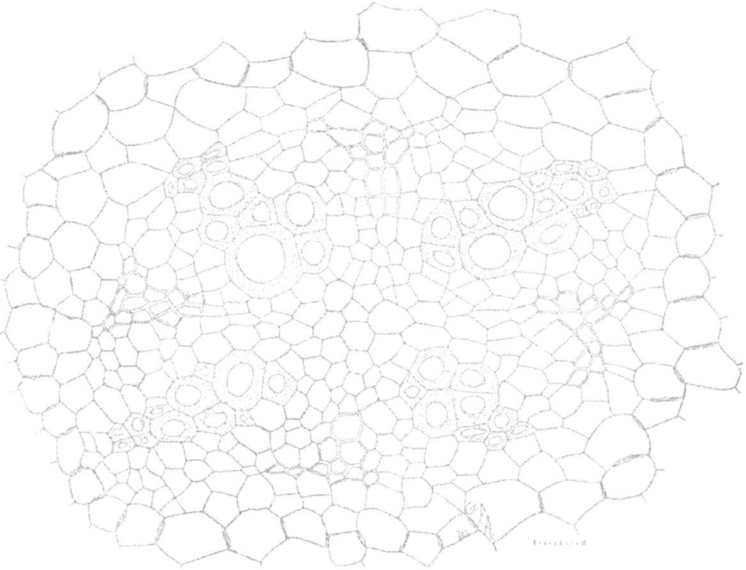

Fig. 163. Transection of the bundle of the root of Botrychium virginianum, a tetrarch bundle in a stout root, showing four groups of xylem alternating with four groups of phloem, and separated by parenchyma. The bundle sheath is the external series of cells shown in the figure. Magnified 30 times more than the scale; scale = 1 mm.

following species have been examined by the author, and in all of them the fungus was present : —

Botrychium matricariæfolium, ternatum, and *virginianum,* from New York State.
B. lanceolatum, Massachusetts and Vermont.
B. subbifoliatum, Hawaiian Islands.
B. lunaria, Cloon Mountains.
Botrychium (No. 484, Drummond's collection, Boston Soc. Nat. Hist.), Louisiana.

Ophioglossum vulgatum, New York State and suburbs of Paris.

O. lusitanicum, Sardinia and Island of Madeira.

O. palmatum, Eastern Cuba.

O. pendulum, Oahu, Hawaiian Islands. The latter was collected on trees, which is quite strong evidence of the probable necessity for the presence of the microsymbiont.

Part II.

METHODS.

NOTES ON TECHNIQUE.

Preparation of Collodion. — The description of methods given here is offered mainly in the way of suggestions upon the technique, such as the writer has found useful, especially in the handling of the more delicate tissues like the prothallia. In connection with this it seemed desirable to state in brief the method of sectioning and preparation of the tissues for study. For full treatment of the technique the student is referred to those works devoted to a wider range of plant histology.

The collodion method employed in the infiltration of the tissues as used here is with some slight modifications adapting it to the delicate prothalline tissue, that described by M. B. Thomas.

Materials. — The materials needed for the preparation of the infiltrating substance are as follows : alcohol — what is known as commercial alcohol with a strength of at least 95 per cent ; ether — the best should be used ; that put up by E. R. Squibbs & Son, Brooklyn, N.Y., can be recommended. Guncotton, as it is known to the trade, is used for the base.

Two solutions of collodion should be prepared as follows. (The quantity of each ingredient given here is that frequently used in making up the solutions. These quantities can be varied according to the quantity of solution desired.)

Two per cent Collodion. — Place 6 grammes of guncotton in a cork-stoppered pint bottle. Pour 150 c.c. 95 per cent alcohol into a graduate, and add to this 150 c.c. of sulphuric ether.

Pour this mixture in the bottle over the 6 grammes guncotton, cork, and as the guncotton dissolves, invert the bottle so that the cork will become wetted with the collodion while it is in place.   This leaves a film of collodion on and around the cork which seals the bottle and prevents the evaporation of the ether, and consequent thickening of the solution.

Five per cent Collodion. — Place 15 grammes of guncotton in a cork-stoppered pint bottle and add the same quantity of a mixture of equal parts of sulphuric ether and 95 per cent alcohol as for the 2 per cent, and when dissolved seal in like manner.

Shake each occasionally for a day or two to aid the solution, when they will be ready for use.   Each time that collodion is used from either solution the cork should be sealed by inverting the bottle, else the collodion will slowly thicken and be unfit for use.

Dehydrating Apparatus.   Tubes. — For dehydrating a few prothallia, 10 to 12, tubes 10 mm. to 12 mm. in diameter and about 6 cm. long are a convenient size.   These can be cut from glass tubing and made in the following way : Cut a circle from good chamois skin 18 mm. in diameter.   Coil elastic wire twice around a tube 8 mm. in diameter, cut off the ends, and bend them inward so they will not cut the skin.   Place the circle of chamois skin over the coil of wire and press it into one end of the tube, being careful not to cut the skin.

Some larger tubes will be necessary for dehydrating the coarser and bulky tissues, and for use occasionally with the smaller ones.   A convenient size for these tubes is $2\frac{1}{2}$ cm. in diameter by 8 cm. to 10 cm. long.   Chamois skin of sufficient size is inserted as described for the smaller tubes.

Dehydrating-jar. — A modified form of Schultze's apparatus can be made as follows : Prepare a circular diaphragm of plaster paris, 18 cm. in diameter by 8 mm. to 9 mm. thick, with perforations of a size to easily admit the large tubes.   Support this diaphragm in a Whitall & Tatum museum jar, about 20 cm. in diameter (in the clear) by 25 cm. in height, so that the diaphragm rests at about two-thirds the height of the jar. While the diaphragm is being made, three glass rods of suffi- cient length to support the diaphragm at the proper height can be inserted as legs.   Pour in sufficient 95 per cent alcohol to reach the diaphragm.   Place rubber bands around the larger dehydrating-tubes so that they can be plunged in the alcohol and be adjusted to any desired level.

Dehydrating and Orienting the Firmer and Older Tissues. — Place segments of the stems, roots, or leaves in 50 per cent alcohol, sufficient to cover them, in a large dehydrating-tube. This is then sunk through a perforation in the diaphragm of the dehydrating-jar to such a depth that the two liquids are at a common level, where it is held in place by adjusting the rubber band.   It is left for twelve to twenty-four hours, according to the tissue.   The 95 per cent alcohol can be kept at the proper strength by adding from time to time some calcium chloride.   Where these tissues of the stem, leaf, and root are young and delicate, it is safer to dehydrate them as is described below for prothallia.

After the tissues are dehydrated they are removed from the alcohol to sufficient 2 per cent collodion to cover them, for about twelve hours, and then in 5 per cent collodion for an equal period.

The segments are then cemented to corks with 5 per cent collodion.   They can be oriented in various positions perpen- dicular to the plane in which it is desired to cut the sections. Where segments are placed perpendicular to the surface of the cork, it may be useful sometimes to cut a shallow hole or slit in the end of the cork.   The 5 per cent collodion can then be placed on with a camel's-hair brush, or poured on in small drops from a vial.   First place a small drop on the surface of the cork, and immediately insert the segment oriented in the de- sired position.   Allow to dry a little.   Drop by drop, or with successive applications with the brush, cover the segment with the collodion, allowing each application to dry somewhat before the next is made. When well covered, allow to dry a very few minutes, then plunge in 80 per cent alcohol. In a few hours it will be ready to section.

Notes on the sectioning and preparation for study are given in the paragraph for the study of the prothallia, and some general comments only will be made here. Sliding cuts should be made with the knife, and it is usually better to draw the knife slowly through the tissues, but sometimes in delicate tissues very thin sections can be made to better advantage by a rather quick stroke of the knife. The usual directions for fixing the sections to the glass slip are to use dry ether vapour, which is blown on the sections from a modified wash bottle. The exit tube does not come in contact with the ether, and is filled part way with broken pieces of calcium chloride to dehydrate the vapour. I have always applied the ether direct, drop by drop, from a dropping-tube. This melts the collodion, and, as the ether evaporates, fixes the sections to the slide, 95 per cent alcohol being floated on as the ether is evaporating. Some objections to the direct application of the ether may be offered. If too much is applied, small sections are apt to float out of their serial order. Also, free parts of sections are apt to float out of position. In such case it may be well to use the ether vapour. In any case the operator might use whichever method is preferred. The free parts of sections are sometimes held in place, while sectioning, by coating the object each time before a cut is made, with 1 per cent collodion, applying it with a camel's-hair brush. The object is then kept dripping with alcohol while being cut.

After the sections are fixed they are washed by floating water over them, stained, washed again with water, dehydrated with alcohol, cleared with the clearing mixture, and mounted in balsam, when they are ready for study.

A clearing mixture which has been found to be useful is made by mixing three parts of turpentine and two parts of melted carbolic acid crystals. If it crystallizes in cold weather, keep it in a warm place while using.

A tray made by cementing glass rods on a rectangular plate, for supporting the glass slips during the process of washing, staining, etc., will be useful.

Two vessels upon a shelf above the work table, one for alcohol, the other for distilled water, with siphon in each for floating alcohol and water upon the preparations, and alcohol upon the knife while cutting, should be provided. Within easy reach the lower end of the siphon, if of glass tubing, can be connected with a delivery jet drawn out from a short section of glass tubing, by a rubber tubing. Upon this section of rubber tubing place a Mohr pinch cock to control the delivery of the liquid. In the alcohol flask the short tube inserted to permit the entrance of air can be partly filled with calcium chloride, so that the moisture in the air will not in the least dilute the alcohol.

GAMETOPHYTIC PHASE.

DEVELOPMENT OF PROTHALLIA.

Sowing the Spores. — Spores of most ferns, the Hymeno-phyllaceæ and Osmundaceæ excepted, require a period of several months' rest before they will germinate. They have been known to germinate when several years old. When possible, for the study of native species, material should be collected in quantity at maturity of the sporangia, and when dehiscence has only begun. The material can be dried in papers in the usual way for herbarium specimens. In lieu of material collected for the purpose, it can be obtained in many cases from the herbarium if not too old. Spores of the Osmundaceæ and Hymenophyllaceæ should be sown as soon as mature, since they possess chlorophyll, and will not germinate after a period of rest and dryness.

Where dried material is not at hand, spores can be obtained from ferns in the conservatories. The material can be sown directly by scraping off the sporangia from the leaf, or the material can be collected in advance at maturity, as described above. If sown directly from the leaves, considerable variability will follow in the germination, as some spores will probably have already passed through a resting period, and will germinate in a few days, while others must yet pass through this period.

The spores should be sown on some porous soil, which can be kept moist from below, but not too wet. The soil can be placed in pots or beds, and be kept covered by a bell jar or

in a Wardian case. The soil should first be sterilized, prefer-
ably by steam or hot water, and the plants will need airing
occasionally. Mr. Shore, head gardener in the botanical conservatories at Cornell University, prepares the soil in the following way : —

The pots used are about four inches in diameter at the top.
Drainage is supplied in the usual way by placing coarse pieces
of cracked pottery in the bottom, covered by finer pieces, the
bits of pottery filling the pots about one-half way. Upon
this is placed a layer of peat moss, followed by a layer of
coarse peat soil sufficient to fill the pot to within one inch
of the rim. This is pressed firmly down. Fine peat soil,
obtained by running peat soil through a sieve, with a little
admixture of fine sand, is then pressed firmly over this about
one-half inch deep. Boiling water is then sprinkled on with
a fine rose watering-pot. The soil is allowed to settle and
cool, and the spores are scattered over the surface. A few
small bits of cracked pottery can be placed about on the soil
before sowing the spores. The spores falling on these afford
clean examples for studying germination and the early development of the prothallia. Too much sand should not be added
to the fine peat soil, since it will hold too much moisture and
endanger the prothallia being killed by "damping off" fungi.
The pots are then placed in a saucer which will hold water,
and all watering is done from below. The pots so prepared
are set away in a box covered with glass to conserve moisture,
some admission of air being provided for, and the box is farther
protected from the sun.

The spores should not be sown too thickly ; for if crowded,
the prothallia develop irregularly, though it would be well to
strew the surface of the soil with them. At the time of germination they can be thinned, and the removal of material for the
study of germination will thin them to some extent. It might
be well to sow some spots very thickly, and by the use of some
loose material have the spores raised somewhat from the sur-

face of the soil. In such cases male prothallia are apt to preponderate, and simple rudimentary protonemal prothallia are developed in profusion, which offer excellent material for the study of the antheridia in a fresh condition.

While waiting for the spores to germinate, the student can take up some other part of the work, but it may be well to study the character of the spores which have been sown. It is also convenient to have some material sown two or three months ahead to provide old prothallia and embryos.

Structure of the Spores. — Mount some of the spores in water. Make descriptive notes and illustrations of their shape; the character of the exospore; the mark of the three-rayed fissure on one side.

Press on the cover glass, moving it about gently, but with sufficient pressure to rupture some of the spores. Note the ruptured exospore; the thin endospore; illustrate.

Stain with iodine; note results.

Germination of the Spores. — In the course of one week to ten days, begin to make examinations of the material for the germinating spores. Tease out some of the spores in water. and cover with a cover glass.

Make descriptive notes and illustrations of the different stages of germination.

The swelling of the spore in germination causes the exospore to rupture. In some cases, especially in *Onoclea*, as Campbell has found, the exospore is entirely cast, and in germination it is thus quite easy to demonstrate the third coat of cellulose which is formed on the spore at this time. The endospore will be seen to split as the rhizoid or first cell of the prothallium protrudes through it.

Note the development of a protonemal thread of a few cells, or, in some cases, a plate of cells. (See text.)

Note contents of cells and the distribution of the various substances.

In very young prothallia note the chlorophyll grains in various stages of division. Search for minute starch grains in the chlorophyll grains. To demonstrate them, place the young prothallia in absolute or 95 per cent alcohol, in the sunlight if necessary, to bleach the chlorophyll grains ; transfer to potassium hydrate, wash with water, and stain with iodine.

Prothallia. — When the material is a little farther grown, if expanded prothallia were not present when studying germination, tease out some of the young prothallia in water, and mount.

Make descriptive notes and illustrations of different stages of development.

Note the expanded plate of cells at one end of the short protonemal thread.

Demonstrate, if possible in young specimens, that this begins and grows for a time usually from a V-shaped, apical cell.

Mount some of the larger prothallia in an inverted position. Note that they are heart-shaped. At the sinus observe the meristematic cushion. Search for archegonia and antheridia and rhizoids. Sketch. (See text for details if not familiar with them.)

Abnormal Prothallia. — From some of the places where the spores were sown very thickly, mount material and search for longer and branched protonemal threads and abnormal prothallia. Sketch and annotate.

If possible, study several different species in the same way, and compare the results.

DEVELOPMENT OF SEXUAL ORGANS.

Selection of Prothallia for Sectioning. — In selecting prothallia for sectioning, great care should be exercised to avoid tearing, bruising, or drying, since the tissue is very delicate. In removing them from the pot where they are grown, it will be well to keep them attached to a small portion of the substratum in a moist chamber ; or, if separated from the substratum, keep them

in a thin film of water in a moist chamber, unless the operator
can take the pot to the table during the selection.

If the prothallia are not growing well separated, place a group
of them upon the glass slip in a thin film of water. With the
needles carefully separate them, placing the points of the
needles among the rhizoids, and directed horizontally along the
prothallia, being very careful not to puncture them. Invert the
prothallia which promise to be good specimens, and examine
them by direct light with a low power of the microscope, to be
certain that they are not torn nor bruised. Search the cushion
of tissue near the sinus for archegonia. They will appear as a
group of small, columnar objects, slightly curved toward the
posterior end of the prothallium, and consisting of four rows of
cells. They will not be readily mistaken for antheridia, since
the latter are quite short. Select only those prothallia which
show quite a number of archegonia. Place the needle under
them, and transfer to the small, dehydrating-tube, which should
be partly filled with water, and resting in a small vessel itself
containing water. Distilled or pure water should be used for
both vessels. In like manner select the number of prothallia
desired. One or two rather young ones, with numbers of
young antheridia, might be selected to insure obtaining the
very young stages.

Now place the dehydrating-tube containing the prothallia in
a small vessel partly filled with 50 per cent alcohol of such a
depth that the liquid in each vessel will be at a common level
at the start. The tube should be supported in such a way that
the bottom does not touch the bottom of the vessel containing
it, in order that there may be free circulation for the liquids.
If possible, agitate the apparatus gently about once every three
hours to assist the prothallia in settling in the liquid. In eight
to twelve hours remove the tube from the 50 per cent alcohol,
and quickly place it in a vessel containing 95 per cent alcohol,
and allow to remain here under like conditions for eight to
twelve hours.

Now pour the prothallia into a flat-bottomed vial with a wide mouth. If they do not flow out at first, pour more alcohol over them from the larger vessel, agitate, and pour quickly into the vial, repeating, if necessary, until all have been transferred. Decant the alcohol from the prothallia, and add sufficient 95 per cent alcohol to cover them well. After one or two hours decant this, allowing it to drain out quite well.

Infiltrating with Collodion. — Add 2 per cent collodion to the prothallia in the vial, sufficient to cover them. Cork the vial, and invert two or three times to seal the cork with collodion from inside. Allow to infiltrate for six hours. Decant the 2 per cent collodion, and add 5 per cent collodion, sufficient to form a thin film of collodion over the bottom of the collodionized paper box when poured in. Cork the vial, and set away in an inverted position for six hours.

Imbedding the Prothallia in Collodion. — Prepare collodionized paper trays as follows : A convenient size is 2 cm. to 3 cm., by 4 cm. to 5 cm., and about 1 cm. deep; but the size must, to a certain extent, depend upon the number of prothallia to be imbedded. Wet the inner surface of the tray with 95 per cent alcohol. Into the bottom of the tray then pour a thin film of 5 per cent collodion, allowing it to dry until it reaches nearly the consistency of hard soap. In like manner add 5 per cent collodion, until the bottom has a layer from 1 mm. to 1.5 mm. in thickness. Pour over the surface 95 per cent alcohol, and allow to stand about fifteen minutes. Decant the alcohol, and allow the wet surface to evaporate until it is just moist, not at all wet.

Now, holding the inverted vial containing the prothallia in 5 per cent collodion over the centre of the tray, remove the cork. With the needles quickly and gently guide the prothallia to rest over the bottom, if possible in an inverted position. Arrange them so that an antero-posterior line through them will run perpendicularly across the tray. This will insure accuracy of blocking and orienting on cork in a position to obtain longitudinal or transverse sections.

Allow the collodion to dry to nearly the consistency of hard
soap, being careful that it does not dry too much on the prothallia, since the layer there is thin. From a vial add more
5 per cent collodion drop by drop, until the prothallia are covered about 1 mm., allowing each addition to thicken as before.
Then nearly fill the tray with 95 per cent alcohol; set aside in
a covered vessel for one or two hours. Decant the alcohol,
strip off the paper from the collodion, avoiding the bending of
the sheet of collodion, which might bruise the prothallia. With
a scalpel cut the collodion in blocks enclosing the prothallia,
by sections parallel and transverse to their antero-posterior axis.

Orienting the Prothallia. — Pour a drop of 5 per cent collodion on the smooth end of a cork. Orient the block containing
the prothallium in this drop, so that the prothallium will stand
on one wing. If the blocks are cut with perpendicular edge,
and in other ways according to directions given above, by thus
cementing one cut edge next a wing of the prothallium, it will
be in a position to get good antero-posterior sections.

If the blocks of collodion do not readily adhere to the fresh
collodion, melt the edge of the block by the application of a
little ether, when the block will adhere without any trouble.

As the collodion in which each block is cemented thickens,
add more 5 per cent collodion, until it is firmly cemented to
the cork. Float the corks in 80 per cent alcohol, and in ten
to twelve hours they will be ready for sectioning.

In dehydrating the prothallia, if the apparatus is properly
prepared, and the prothallia are agitated about every one to two
hours, six hours in 50 per cent alcohol and the same length of
time in the 95 per cent alcohol, with one to two hours in 95 per
cent alcohol when transferred to the vial, will be ample time.
But, on the whole, it may be safer to extend the time somewhat. Frequently good results are obtained by lowering the
small dehydrating-tube containing the prothallia with pure
water, into 50 per cent alcohol in a larger tube, which in turn
is lowered into the dehydrating-jar containing 95 per cent

alcohol. After twenty-four hours, having agitated the apparatus a few times, pour into the infiltrating-vial, and allow to
remain for an hour or two in fresh 95 per cent alcohol, and
proceed as before. The smaller dehydrating-tube should not
rest nearer than one-half inch from the membrane in the larger
tube, and the liquids in all cases should be at a common level
at the start.

Sectioning Prothallia. — For antero-posterior sections, fasten
a cork with its prothallium in the jaws of the microtome, so
that the cut will pass perpendicular to the surface of the prothallium at the centre, and parallel with the antero-posterior
axis. Trim off the surplus collodion around the object.

Unless for some special reason it is desirable to save serial
sections of the entire prothallium, none need be saved except
in the region of the archegonia. In making sections through
this region, usually a sufficient number of antheridia will be
secured for study. If this should not be the case, section one
or two prothallia having a number of young antheridia.

The sections should be made with some instrument of pre-
cision like a microtome, in order to obtain uniformity of thickness, and to save with as great a degree of certainty as possible
all the sections. The knife should be very sharp ; a good razor
is excellent. The tissues are very delicate and easily torn.
Keep sufficient alcohol on the knife to float the sections, in
order that they may not tear nor fold. With the knife at an
angle of thirty to forty-five degrees with the stroke, start the
sections near the edge of the prothallium. Cut with a steady
and rather quick stroke, but not with a jerk. With a needle
or camel's-hair brush lead each section at a short distance on
the blade, so that successive ones will not fold upon them.
Having cut three or four sections, float them on a glass slip,
and while wet with alcohol examine them with the low-power
microscope. If the region of the archegonia has not been
reached, discard them and cut a few more, and so proceed until
the sections begin to enter the cushion bearing the archegonia.

If the prothallium can be seen in the block, the sections can be made quite thick through the wing until near the meristematic cushion. At any rate, they can be made much thicker while approaching the cushion. A little experience will be necessary to determine the thickness of the sections, since some prothallia are coarser than others, and the sexual organs are apt to vary in a corresponding degree.

Through the cushion bearing the archegonia my sections vary from 25 μ to 50 μ; more frequently they are about 40 μ in thickness.

Having transferred to the glass slip the number of sections desired, or such a number as can be conveniently covered with the cover glass used, proceed to arrange and fix them. Having kept them in serial order while transferring them to the glass slip, arrange them as closely as desired, so that they will read in successive lines, either transversely or longitudinally with the slip, some uniformity being preserved in this matter in order to avoid confusion at a later period. During the entire process of transfer and arrangement they must be kept wet with alcohol, and should not be allowed to dry at any subsequent stage.

Fixing the Sections to the Glass Slip. — Now permit the alcohol to evaporate (or draw it off with absorbent paper) until the sections begin to dry. Then with a dropping-tube place on a drop of ether to melt the collodion, and fix the sections to the slip. The ether will quickly evaporate, but before the sections become dry and white float on alcohol. In a few moments wash with water.

Stain with hæmatoxylin; wash again with water.

Float on 95 per cent alcohol for fifteen to thirty minutes, followed by the clearing mixture for an equal time. Remove excess of clearing mixture and mount in balsam.

If trouble is experienced in preventing the sections from floating far out of position, a less amount of ether can be added at one time, followed by successive applications just as the previous one is disappearing, until the collodion is sufficiently melted and the sections are fixed. It is sometimes advantageous to fix the sections, three or four at a time, as they are transferred from the knife to the glass slip, especially where a large number of serial sections are to be fixed for a single mount.

Immediately after fixing with ether, when covered with alcohol, if the sections or the collodion still on the slip should appear whitened or cloudy, add a drop of ether while the objects are wet with alcohol. Then quickly float on more alcohol just as the cloudiness is disappearing, in order to prevent the ether from loosening the sections again.

Experience will determine the length of the time for the working of the stain, also for dehydrating and clearing the stained objects. To hasten the dehydration, the first application of alcohol can be poured off in a few minutes and fresh alcohol be floated on a second time. The clearing can also be accelerated in the same way.

Should cloudiness appear after mounting in balsam, there has been some want of care at some stage of the process. Sometimes this can be remedied by removing the cover immediately and adding fresh clearing mixture again.

Archegonia and Antheridia. — (In connection with this see text descriptive of the sexual organs.) In the sections from a single prothallium, if care was used in the selection, several different stages of development of the archegonia and antheridia will be found. In studying the sections, search for the very earliest stages and note the succession of cell division as the organs become more mature.

Make careful descriptive notes and drawings of various stages of young antheridia and archegonia; stages of division of the central cell and spermatozoids in the antheridia.

The central cell, ventral, and neck canal cells of the archegonia.

Study open antheridia and open archegonia.

Search for various stages in fertilization, spermatozoids caught in the slimy protoplasm of the canal cells at the mouth of the archegonia and in the canal.

Several prothallia should be prepared and examined in order to get a good representation of the different stages.

Search for different stages of the rhizoids.

Study the structure of the prothallia in section.

Antheridia in a living condition can be studied by mounting simple male prothallia in water. The rupture of the cap cell and expulsion of the spermatozoids should be noted ; note also the quiet condition of the spermatozoids at first, but that soon they whirl away. Usually it requires a good homogeneous immersion lens to see the cilia.

For the study of fertilization it will be well to fix the material after the selection of the prothallia in a 1 per cent solution of chromic acid. The prothallia should be then soaked in water to remove the dark stain ; then dehydrated, infiltrated with collodion, and prepared for study in the usual way.

After dehydration it may also be convenient sometimes to stain the prothallia in bulk before the infiltration with collodion.

DEVELOPMENT OF THE EMBRYO.

Selection of Prothallia. — The prothallia should be examined for a small protuberance from the under side of the meristematic cushion. Usually accompanying this will be an increased development of rhizoids from this region. A few prothallia should also be selected, showing the very young cotyledon rising to pass through the sinus at the anterior end. To observe the very early stages of the embryo, select prothallia with old archegonia, upon which there is no external appearance suggesting the embryo.

These should be prepared for study the same as described above. Make notes and drawings of as full a series of the embryo as can be obtained. Note the basal and transversal walls, if such young stages are obtained, dividing the embryo into the stem, leaf, root, and foot segments ; the apical cell of stem and leaf ; apical cell and root cap of young root. (The text on the development of the embryo may be consulted here.)

In preparing sections for the study of the archegonia, young embryos will sometimes be obtained, and also good archegonia will frequently be obtained in sectioning for embryos.

Cases of apogamy should be looked for. They can be easily obtained by sowing the spores of *Pteris cretica*.

SPOROPHYTIC PHASE.

MACROSCOPIC STUDY.

For the general morphology of the sporophyte the choice may fall on almost any of the medium-sized ferns. The ease of obtaining material in good condition should govern the selection rather than the insistence on any one species. There are several species, however, which, both for their wide distribution and for their convenience in collecting, can be recommended for study. Of these, *Pteris aquilina, Adiantum pedatum, Polypodium vulgare, Onoclea sensibilis,* or their near relatives are convenient for the study of the stem, since the leaves are not crowded. Various specimens of *Aspidium* or *Asplenium* provide material for the study of crowded leaves representing a more complicated phyllotaxy.

Wherever possible, several of these should be obtained in order to make a comparative study. With this in view species of *Lygodium, Schizæa,* or *Hymenophyllum,* offer the single central bundle which is interesting to compare with the other arrangements found to prevail in the species mentioned above. Species of *Botrychium* or *Ophioglossum* will also repay study.

The plants should be removed from the substratum with care to avoid injuries to the bud or growing end, and to preserve as many roots free from injury as possible, especially the root tips. If it is necessary to collect them some time in advance, the parts may be preserved in 80 per cent alcohol, having first dehydrated them by passing successively through 50 per cent, 75 per cent, and 95 per cent alcohol; but it is

always better when possible, especially for microscopic study
of the more delicate tissues, to collect and prepare for use as
indicated under the paragraph on microscopic study.

The Stem. — Descriptive and comparative notes, and drawings when useful, should be made of the general morphology
of the stem ; the branching ; terminal bud or growing end ;
the dying end ; peculiarities of form ; protective covering of
certain parts ; leaf scars, or bases of attached leaves.

Root. — Observe the succession of the roots, character of
their branching, position of the root hairs, the root tip.

Leaf. — (Herbarium material will answer for macroscopic
study.) Note the attachment of the leaves to the stem ; compare the relative position and number in different species. In
such forms as *Aspidium acrostichoides, Asplenium angustifolium, Onoclea, sensibilis,* etc., note the greater or lesser differentiation of the fertile and sterile leaves. If conservatories are
convenient, note the differentiation of leaves in the case of
some of the epiphytic ferns. (See text.)

Parts of the Leaf. — The stalk and lamina, the latter usually
farther differentiated into the rachis and pinnæ. Sketch.

The Stalk. — Note form ; compare extent with that of the
lamina ; observe character of the surface, colour, texture.

The Lamina. — From herbarium material, note the simple
character of the lamina in such ferns as *Scolopendrium vulgare*
and *Camptosorus rhizophyllus.* In other ferns, compare the
various degrees of complexity in the division of the lamina into
rachis and pinnæ, and the various forms presented. Sketch
forms of pinnæ and pinnules, and note carefully the character
of the venation.

The Reproductive Bodies. — On fertile leaves, or fertile parts
of leaves, observe the "fruit dots," or sori. From herbarium
material, if fresh material is not at hand, compare position,
form, exposure, or covering (indusium), of the sori of several
different species. (Consult text.) Sketch various forms, and
make comparative annotations ; with a pocket lens observe the

stalked sporangia forming the sorus, and the form of the indusium when present.

The Stem. — To prepare stems for sectioning, cut several segments about 6 mm. long containing the bud or growing end ; without injuring the bud, mark the base of the segment in some way, so that the perpendicular (to ground when the stem is in natural position) and horizontal diameter can be determined in orienting the segment for sectioning.

Dehydrate and infiltrate with collodion. It will require a longer time to dehydrate the stems than the thinner and more tender leaf and prothalline tissue. Twelve hours in the 50 per cent alcohol, and an equal time in the 95 per cent, or the older parts of the stem may be put directly in the 50 per cent alcohol in the dehydrating-tube, and this sunk in the 95 per cent alcohol for twelve to twenty hours. If it is necessary to collect the material some time before it can be studied, these segments, after having been dehydrated, can be allowed to stand in 80 per cent alcohol, or they can be carried through the collodion, oriented on corks, and stored in 80 per cent alcohol, until ready for use. As a last precaution, I have found it well to place the material in a small quantity of fresh 95 per cent alcohol for an hour or so before placing in the 2 per cent collodion. If the material is stored in 80 per cent alcohol after dehydrating, when ready for later use it must be again placed in 95 per cent before placing in the collodion. Allow the segments of the stem to remain in the 2 per cent and 5 per cent collodion twelve hours respectively. The segments of the stem, being convenient to handle without injury, can be oriented on the corks directly. Some of the shorter ones can be oriented in a horizontal position, segments of the buds, or growing ends, being so placed that some will section in a longitudinal horizontal plane ; others in a longitudinal vertical plane, to obtain different views of the apical cell. To orient the segments for transverse sections, cut a shallow hole in the end of a cork to receive one end. Segments of the buds should be similarly mounted in order to obtain transverse sections of the apical cell. The segments can be cemented to the corks in the same manner as described for the collodion blocks containing prothallia.

Cutting the Sections. — Cut the sections of the various specimens, fix with ether, harden with alcohol, wash with water, stain with hæmatoxylin, wash with water, dehydrate with alcohol, clear with the clearing mixture, and mount in balsam.

In longitudinal sections of the bud, serial sections should be carefully saved through the middle in order to obtain the apical cell. In transverse sections of the bud, serial sections must be carefully saved from the very first entrance into the tissue until the apical cell shall have been passed.

Longitudinal sections of the older portions of the stem should contain the greatest diameter of some of the bundles. In specimens like *Botrychium* or *Adiantum* a long and complete series of transections should be preserved, showing the transition from one leaf trace to another. Quite a long series can be preserved on a single glass slip by using long rectangular cover glasses.

In *Pteris, Polypodium, Onoclea*, etc., a few transections will suffice.

Study. (See text for reference.)

Apical Cell. — From longitudinal and transverse sections of the bud make out the form of the apical cell.

Note the successive oblique divisions on each side of the apical cell; the stratified condition of the tissues for some distance ; the outline of the end of the bud in longisection ; the protective scales.

In such forms as *Adiantum, Onoclea, Botrychium*, etc., note the young leaves coiled over the bud. Search for very young leaves, shown by the arching of tissue near the apical cell in longisection. Frequently the apical cell of the leaves can be seen in these sections. Look out for longitudinal and transverse sections of young roots in various stages of development, not yet emerged from the stem. They can be determined by the apical cell and root cap. Note the change of form in cells at varying distances from the apical cell. Some elongate, gradually lose their protoplasm, and become tracheides of the vascular bundles.

Trace the connections of these transforming cells in stem, leaf, and root. Sketch the various features and make full annotations.

Structure of the Stem. Transection. — With the low power of the microscope study the arrangement of the various tissues. The things to be observed are —

Sclerenchyma, dark brown tissue, with thick cell walls. Note its arrangement. If several species are examined, note its varying distribution and occurrence.

Vascular Bundles, groups of large, empty cells with stout walls (xylem), surrounded by smaller cells, usually with protoplasmic contents (phloem), this surrounded by a chain of cells (bundle sheath). Such a bundle is concentric.

Parenchyma. — Thin-walled, large cells, with prominent interspaces.

Leaf Traces and Origin of Roots. — These are beautifully shown in a series of sections of the stem of *Adiantum pedatum*. The leaf trace, C-shaped, opens against a larger C-shaped bundle in the stem, while the root bundles fork off from the outside of the stem bundle without opening it. Study the changing form of the bundle in a series of sections.

Starch Grains, in cells of parenchyma and sclerenchyma.

Sketch in outline the grouping of tissues in transection of the stem. Annotate.

With the high power of the microscope study the markings and other characters of the thick-walled sclerenchyma.

Vascular Bundles. — In the centre of the bundle note the character of the thick-walled xylem cells (tracheides); the thin-walled cells (parenchyma) intermingled with them; outside of this xylem a ring of smaller cells with slightly thickened walls, which stain quite deeply with hæmatoxylin (bast), intermingled with larger thin-walled cells (parenchyma), the whole forming the phloem portion of the bundle. Surrounding the bundle note a chain of usually narrow cells, the bundle sheath; just inside of this a row of larger cells, the phloem sheath.

Longisection. Vascular Bundle. — Detect the various tissues studied in transection. Certain markings on the cell walls enable us to differentiate them into —

Scalariform Tracheides, elongated cells with transverse pits. Where two longitudinal walls meet, note the bordered pits.

Spiral Tracheides, usually smaller cells with spiral markings.

Scalariform Vessels. — In *Pteris aquilina* certain of the oblique walls connecting the ends of the cells are perforated by pits, forming true vessels.

Xylem Parenchyma, thin-walled parenchyma cells intermingled with the xylem.

Sieve Tubes, elongated cells bordering the tracheides with sieve plates in their walls.

Bast, narrow, elongated cells bordering usually the sieve tubes.

Phloem Parenchyma, large cells accompanying the bast.

Phloem Sheath and Bundle Sheath in succession.

Sketch a bundle in longisection, showing details of structure. If possible, compare stems of *Botrychium* and *Ophioglossum.* The bundles here are collateral. (See text for detail.)

Root. Selection of Material. — Short segments 4 mm. to 6 mm. from the larger roots, and segments containing the root tip of almost any species, can be used. If possible, roots of *Botrychium* and *Ophioglossum* should also be provided. While the root cap is not so highly developed in these as in the ferns, the root and tip are large and easily prepared for study, and serve as excellent material for comparison. The material can be prepared for study as described for the stems.

Apical Cell. — From longisections and transections of the root tip, study the form of the apical cell; the succession of cells cut off on all sides; the stratification of the cells; the transition of these cells into the various tissues; note the root hairs, and compare with the Ophioglosseæ. Sketch the study and annotate.

Structure of Root. Transection. — Note the epidermis; cortical parenchyma, sometimes brown; sclerenchyma inside this forming a ring around the single, radial, vascular bundle; bundle sheath; xylem; scalariform tracheides, usually arranged in two opposite groups, which in old roots are sometimes connected. In younger roots note very large, thin-walled cells in the centre separating the two xylem groups. These in age change to tracheides, thus connecting the two groups.

Phloem, alternating with the xylem groups; note soft, thick-walled cells which take a deep stain from the hæmatoxylin.

Parenchyma. — Between the xylem and phloem. This is gradually changed to xylem and phloem in age.

Sketch and annotate the study.

Leaves. Preparation of Material. — The structure of the leaf can be studied from the same material which is used for the development of the sporangia, though it may be well to put up some material representing sterile parts of the lamina.

Material to be used for the study of the development of the sporangia and spores should be selected from fertile leaves; some from material bearing very young sori, and some with sori bearing sporangia in the final stages of development.

Dehydrate with alcohol, and infiltrate with collodion as described for the prothallia.

Orient on corks in such a way as to obtain transections of the leaf through the sorus. Portions of the sterile leaves can be oriented so as to obtain both longitudinal and transverse sections. Section and prepare for study.

Structure of Leaf. — In transections of the leaf, descriptive notes and drawings should be made of the structure of the leaf, showing mesophyll, epidermis, and vascular bundles.

Epidermis. — Surface views of the epidermis can be best studied from fresh material. With fine-pointed forceps pick the lower epidermis, and tear off a strip and mount in water. Note irregular form of cells; the stoma and guard cells; chlorophyll in the epidermal cells. Stain with iodine. Sketch and annotate.

Prepare a strip from the upper epidermis in the same way. Sketch and annotate.

Development of the Stomates. — Excellent preparations showing the developmental stages of the stomates can be prepared as follows : Select terminal portions of the young leaves which are just unrolling. (I used *Pteris aquilina*.) Cut a segment 4 mm. to 6 mm. long from the ends of several leaves. Dehydrate and infiltrate in the usual way. To orient, allow a drop of collodion to stiffen on the end of a cork. Place the young leaf segment horizontally on this with the lower epidermis uppermost. Section and prepare for study. Though it will be wellnigh impossible to get the entire epidermis, or any very large portion of it, in a single plane, at the edges of many of the sections will be found considerable portions of it.

Observe that the epidermal cells are much smaller than in the mature leaf, and that their form is more simple. Note semicircular walls in the side of some, forming mother cells for the guard cells. Note various stages of development of the guard cells ; the abundance of protoplasm in the young epidermal cells and large nuclei in different stages of division. (Consult text.) Sketch and annotate.

Development of Sporangia and Spores. — (Consult text.) Note placental region ; its proximity to a vascular bundle. In sections of the younger sori search for the earliest origin of the sporangia as epidermal outgrowths from the placental region. By searching among the different young sporangia trace the successive cell divisions to produce young sporangial wall ; tetrahedral archesporium ; tapetal cells ; tetrahedral primordial mother cell of the spores ; successive bipartitions of this into

sixteen mother cells of the spores with large nuclei ; the dissolution of the tapetal cells.

Spores. — Search for mother cells, the nuclei of which show
the nuclear spindle, and various stages of its formation and
division producing primary bipartition of the nucleus. Search
for mother cells showing the farther division of the nuclei into
four nuclei. Search for successive stages in the farther development of the spores and formation of the walls. Sketch and
annotate.

In mounting dry sporangia the collodion should be allowed to
dry quite hard at each application, so that the objects will not
pull out of position by the knife.

Dehiscence of Sporangia. — If recently matured sporangia
can be procured, proceed in the order of paragraphs from 1 to
10. If not, then with dry material proceed in the following
order of paragraphs : 7, 1, 8, 9, 3, 4, 5, 6, 10.

1. *Polypodiaceæ.* — To note the first dehiscence and dispersion of the spores, specimens should be selected the sporangia
of which are mature but not yet open. Take a pinnule or
portion of leaf of any species the sporangia of which are free
or project from underneath the indusium. Invert upon a glass
slip and examine with a low power of the microscope by direct
light. If movement does not promptly begin, hold the slip a
moment over a low flame and quickly return to the stage of the
microscope. Note the slow elevation of the distal end of the
annulus as it straightens, and that the mass of spores is held
near the end between adherent portions of the lateral walls.
Keep the attention fixed upon the annulus until it is completely
everted, the ends usually meeting, when with a sudden flip it
scatters the spores and returns to nearly its former position.

In some of the sporangia note that before the annulus is
completely everted it snaps for a short distance, at the same
time continuing to evert, until by several successive snaps it
eventually comes to rest somewhere between a nearly straight
line and the closed position.

2. Now scrape some of the mature unopen sporangia from the leaf to the glass slip. In a dry condition cover with a cover glass and examine with the low power of the microscope by transmitted light, heating very gently if necessary. When the annulus begins to move, note that the rift in the sporangium begins at the stomium between the lip cells, and continues, after parting them, irregularly across the lateral walls. The movement can now be more easily observed, but the cover interferes somewhat with the dispersion of the spores. Sketch different positions of the annulus and sporangium.

3. After movement has ceased, run water under the cover glass and note quickly that within each cell of the annulus of the dehiscent sporangia is a sphere of air. Focus the high power on some of these and note that the spheres gradually grow smaller until, with a sudden whiff, they one by one vanish, having been absorbed by the water under endosmotic pressure. Sketch different appearances.

4. Now place a small drop of glycerine on the slip by the side of the cover and with absorbent paper draw off the water. As the stream of glycerine comes in contact with the sporangia, note that the effect on the annulus is the same as that produced by drying. After the annulus has sprung, note that each cell is again occupied by a sphere of air.

5. Remove the cover and with a needle move the sporangia to one edge of the glycerine; wash them once or twice with water quickly. Add fresh water and cover with the cover glass. Observe the disappearance of the air spheres again. Add glycerine a second time and note results.

6. With fresh material, instead of glycerine, use alcohol, or strong solutions of potash, chloriodide of zinc, or other substances which absorb water.

7. *Dry Material.* — (Material should be selected which is mature, but which has not weathered by long exposure to weather after maturity.) From the herbarium or from material collected after the sporangia have opened take a small portion

of leaf or pinnule, and after moistening the sporangia with water invert on a glass slip. Dry by gentle heat, and note movement of the annulus as described in paragraph 1.

8. Scrape some dry sporangia to the glass slip, moisten with water, aiding the absorption of the water by gentle pressure from a scalpel or spatula. Now by gentle heat evaporate the water, and just as the moisture is disappearing examine with the low power.

9. Make a similar preparation, but after adding the water, place on a cover glass. Now quickly note the air spheres in the cells of the annulus, and proceed as directed in paragraphs 3, 4, 5, 6, and 10.

10. Try material from very old herbarium specimens.

BIBLIOGRAPHY.

—◆—

THE following bibliography includes the more important papers and works consulted in connection with the study. Systematic works are not included, since they bear a less important relation to the present subject than those on development.

ATKINSON, GEO. F., Symbiosis in the Roots of the Ophioglosseæ; Bulletin Torrey Botanical Club, XX., 1893, p. 356.

—— Unequal Segmentation and its Significance in the Primary Division of the Embryo of Ferns; Bull. Torr. Bot. Club, XX., 1893, p. 405.

—— Two Perfectly Developed Embryos on a Single Prothallium of Adiantum cuneatum; Bull. Torr. Bot. Club, XX., 1893, p. 407.

—— The Extent of the Annulus, and the Function of the Different Parts of the Sporangium of Ferns in the Dispersion of Spores; Bull. Torr. Bot. Club, XX., 1893, p. 435.

DE BARY, A., Comparative Anatomy of the Phanerogams and Ferns, 1887 (English translation).

—— Ueber die von Farlow zuerst beschriebene Bildung beblätterter Sprosse an Farn-Prothallien; Tageblatt der 50 Versammlung deutscher Naturforscher und Aertze, p. 200.

BAUKE, H., Entwickelungsgeschichte des Prothalliums bei den Cyatheaceen vergleichen mit derselben bei den andern Farnkräutern; Jahrb. f. wiss. Bot., X., p. 49 (J. J. B., V., 1887, p. 282).

BEHRENS, J. W., A Guide to the Microscope in Botany; 1885. English edition.

BENNETT AND MURRAY, Cryptogamic Botany; 1889.

BERGGREN, Ueber Apogamy des Prothalliums von Notochlæna distans; Bot. Centralb., XXXV., p. 183.

BESSEY, C. E., Botany.

BOWER, F. O., Preliminary Note on the Formation of Gemmæ on Trichomanes alatum; Ann. Bot., I., 1887–88, p. 183.

—— On Some Normal and Abnormal Developments of the Oophyte in Trichomanes; Ann. Bot., I., 1887–88, p. 269.

—— On Apospory and Allied Phenomena; Trans. Linn. Soc., London, 2 ser. Bot., II., 1887, p. 301.

BOWER, F. O., Comparative Examination of the Meristems of Ferns as a Phylogenetic Study; Ann. Bot., III., No. 2.

CAMPBELL, D. H., A Third Coat in the Spores of the Genus Onoclea; Bull. Torr. Bot. Club, XII., 1885, p. 8.

—— The Development of the Prothallia of Ferns; Bull. Torr. Bot. Club, XII., 1885, p. 355.

—— Development of the Root in Botrychium ternatum; Bot. Gazette, XI., 1886. p. 49.

—— The Development of the Antheridium in Ferns; Bull. Torr. Bot. Club, XIII., 1886, p. 49.

—— The Development of the Ostrich Fern; Mem. Boston Soc. Nat. Hist., IV., 1887, p. 17.

—— Zur Entwicklungsgeschichte der Spermatozoiden; Ber. d. deutsch. Bot. Gesells., V., 1887, p. 120.

—— On the Affinities of the Filicineæ; Bot. Gaz., XV., 1890, p. 1.

—— Notes on the Apical Growth of Osmunda and Botrychium; Bot. Gaz., XVI., 1891, p. 37.

—— A Study of the Apical Growth of Ferns with Reference to their Relationships; Bull. Torr. Bot. Club, XVIII., 1891, p. 73.

—— Notes on the Archegonia of Ferns; Bull. Torr. Bot. Club, XVIII., 1891, p. 16.

—— On the Relationships of the Archegoniata; Bot. Gaz., XVI., 1891, p. 323.

—— On the Prothallium and Embryo of Osmunda claytoniana and O. cinnamomea; Ann. Bot., VI., 1892, p. 49.

CRAMER, Ueber die geschlechtlose Vermehrung des Farn-prothalliums; Denkschr. Schweiz. Naturf. Gesells., XXVIII., 1880.

DRURY, C. F., Observations on a Singular Mode of Development in the Lady Fern; Jour. Linn. Soc., London, XXI., p. 354.

—— Further Notes on a Singular Mode of Reproduction in Athyrium filix-fœmina, var. clarissima; J. L. S. L., XXI., p. 358.

FARLOW, W. G., Asexual Growth from the Prothallium of Pteris cretica; Quar. Jour. Micr. Soc., 1876, p. 266.

—— Apospory in Pteris aquilina; Ann. Bot., II., 1888, p. 383.

GOEBEL, K., Entwickelungsgeschichte des Prothalliums von Gymnogramme leptophylla; Bot. Zeit., 1877, Nos. 42-44.

—— Zur Embryologie der Archegoniaten; Arb. d. Bot. Inst. in Wurzburg, II., 1880, S. 437.

—— Morphologische und Histologische Notizen; Ann. d. Jard. Bot. d. Buitenzorg, VII., 1887, S. I. (J. J..B., XV., I., 563).

—— Outlines of Classification and Special Morphology of Plants.

HEINRICHER, Beeinflusst das Licht die Organ Auflage am Farn Embryo? Mitth. d. Bot. Inst. z. Graz, II., S. 239.

HOFFMEISTER, Higher Cryptogamia; 1862.

HOLTZMAN, C. L., On the Apical Growth of the Stem and the Development of the Sporangium of Botrychium virginianum; Bot. Gaz., XVII., 1892, p. 214.

KIENITZ-GERLOFF, F., Ueber den genetischen Zusammenhang der Moose mit den Gefässkryptogamen und Phanerogamen; Bot. Zeit., 1876, Nos. 45–46.

—— Untersuchungen über die Entwickelungsgeschichte der Laubmooskapsel und die Embryoentwickelung einiger Polypodieen; Bot. Zeit., 1878, Nos. 3–4 (J. J. B., VI., I., 1878, 535).

KNY, L., Keimungs- und Entwickelungsgeschichte von Ceratopteris; Nova Acta d. K. Leop. Carol. Deutsch. Akad. d. Naturf., XXXVII. (J. J. B., 1874, 385).

—— Die Entwickelung der Parkeriaceen, dargestellt an Ceratopteris thalictroides; Ibid. (J. J. B., 1875, 333).

KÜNDIG, J., Beiträge zur Entwicklungsgeschichte der Polypodiaceen Sporangium; Hedwigia, 1888, p. 1.

LECLERC DU SABLON, Recherches sur la dissemination des Spores chez les cryptogames Vasculaires; Ann. d. Sc. Nat. Bot., 7th ser., Tom. II., 1885, p. 5.

LEITGEB, Zur Embryologie der Farne; Sitzungsb. d. Mathem. Klasse d. k. Akad. d. Wiss. zu Wien, LXXV., 1878 (J. J. B., VI., I., 5).

—— Ueber Bilateralität der Prothallien; Flora, 1879, S. 317 (J. J. B., VII., I., 409).

LYON, F. M., Dehiscence of the Sporangium of Adiantum pedatum; Bull. Torr. Bot. Club, XIV., 1887, p. 180.

PRANTL, K., Die Mechanik des Rings am Farn-Sporangium; Tagebl. d. 52 Versaml. deutsch. Naturforscher u. Aertze in Baden Baden, 1879, S. 213.

—— Ibid.; ber. d. Deutsch. Bot. Gesells., IV., 1886–87, S. 41.

—— Ueber den Einfluss des Lichtes auf die Bilateralität der Farnprothallien; Bot. Zeit., 1879, S. 697 and 713.

RAUWENHOFF, De Geslachtsgeneratie der Gleicheniaceen; Verh. d. konikl. Akad. van Wetensch. te Amsterdam, 1889 (J. J. B., XVII., I., 718).

REES, Sporangien; Pringsheim's Jahrb. Wiss. Bot., 1867, p. 217.

SACHS, The Physiology of Plants.

SCHINZ, Untersuchungen über den Mechanismus des Aufspringens der Sporangien und Pollensäcke; Zurich, 1883.

SCHRENK, JOSEPH, Dehiscence of Fern Sporangia; Bull. Torr. Bot. Club, XIII., 1886, p. 168.

SCHRODT, Das Farnsporangium und die Anthere; Flora, 1885.

—— Der mechanische Apparat zur Verbreitung der Farnsporen; Ber. d. Deutsch. Bot. Gesells., III., 1885–86, p. 396.

—— Neue Beiträge zur Mechanick der Farnsporangien; Flora, 1870, p. 70.

SEDGEWICK AND WILSON, Biology; 1886.

SMITH, W., Longevity of Spores; Gard. Chron., 1889, II., p. 140.

SMITH, J., Historia Filicum; 1875.

STRASBURGER, E., Ueber Befruchtung und Zelltheilung; 1878.

—— Botanisches Practicum, 1884.

—— Histologische Beiträge, I.–IV.

— — AND HILLHOUSE, Handbook of Practical Botany, 1887.

THOMAS, M. B., The Collodion Method in Botany; Bot. Gaz., XV., 1890.

— · · - Ibid.; Proceed. Am. Soc. Microscp., 1890. p. 123.

UNDERWOOD, L. M., Our Native Ferns and their Allies.

— Onoclea sensibilis, var. obtusiloba; Bull. Torr. Bot. Club, VIII., 1881, p. 101.

VINES, S., On the Homologies of the Suspensor; Quar. Micros. Jour., 1877, p. 58.

VOUK, F., Die Entwickelung des Embryo von Asplenium shepherdi; Sitzungs b. d. k. k. Akad. d. Wiss., LXXVI., I., 1877 (J. J. B., V., p. 280).

WORKS ON BOTANY, ETC.

PUBLISHED BY

MACMILLAN & CO.

————————

ALLEN (GRANT).—**On the Colours of Flowers.** With Illustrations. *Nature Series.* $1.00.

"... A copiously illustrated work by Grant Allen which cannot fail to interest students of evolution. Mr. Darwin said of it that it was an evolutionary argument 'too striking and too apparently valid ever to be forgotten.' Perhaps we can best indicate the extent and force of this little book by giving the suggestive words with which it closes: 'If the general principle here put forward is true, the special colors of different flowers are due to no mere spontaneous accident; nay, even to no meaningless caprice of the fertilizing insects. ... Not only can we say why such a color, once happening to appear, has been favored in the struggle for existence, but also why that color should ever make its appearance in the first place, which is a condition precedent to its being favored or selected at all.' To attempt merely an outline of this theory would be to do its logical and continuous strength a grave injustice. We can only refer those who desire such a knowledge to Mr. Allen's own felicitous and scientific explanations. Certainly it constitutes a valuable acquisition to the 'Nature Series' already favorably recognized."— *Chicago Tribune.*

AVELING. — **An Introduction to the Study of Botany.** 271 Illustrations, and a Glossary of over 600 words. $1.10.

"After a short general introduction, some two hundred pages are devoted to the dissection and description of eighteen or twenty common plants, such as the buttercup, orpine, hyacinth, pea, primrose, dandelion, daisy, foxglove, hemlock, etc. Every part of the plant is minutely described, and many illustrations and diagrams accompany the text. ... The next ninety pages are devoted to structural botany, beginning with the vegetable cell, and so developing the subject to the fruit. ... Throughout the book, a Latin or Greek derivation is generally given to technical terms."— *New York Evening Post.*

BAKER (J. G.). — **Handbook of the Irideæ.** 8vo. $1.75.

Handbook of the Amaryllideæ, including the **Alstrœmerieæ** and **Agareæ.** 8vo. $1.75.

BALFOUR (J. HUTTON). — **Manual of Botany:** an Introduction to the Study of the Structure, Physiology, and Classification of Plants. With Glossary of Botanical Terms and Index. Fifth edition, illustrated with 963 Woodcuts. 12mo. $3.00.

1

BALFOUR (J. HUTTON). — *Continued.*

> **Elements of Botany,** for the Use of Schools, with a Glossary of
> Botanical Terms and Index.   With 427 Woodcuts.   16mo,
> cloth. $1.00.

BELL'S AGRICULTURAL SERIES. — In 12mo.   Illustrated, cloth.
> 75 cents each.

> **The Farm and the Dairy.**   By Prof. J. P. SHELDON.

> **Soils and their Properties.**   By Dr. WILLIAM FREAM, B.Sc.

> **Manures and their Uses.**   By Dr. A. B. GRIFFITHS.

> **The Diseases of Crops and their Remedies.**   By Dr. A. B. GRIFFITHS.

> **Tillage and Implements.**   By W. J. MALDEN.   (*In the Press.*)

> **Practical Fruit-culture.**   By J. CHEAL, F.R.H.S.

> **Feeding Stock.**   By HENRY EVERSHED.   (*In the Press.*)

> **Veterinary Medicine.**   By Prof. W. F. GARSIDE and J. H. DUGDALE.
> (*In the Press.*)

BETTANY (G. T.). — **First Lessons in Practical Botany.**   18mo.
> 30 cents.

BLOMFIELD (R.) and **THOMAS** (F. I.). — **The Formal Garden in
England.**   With numerous Illustrations.   Crown 8vo. $3.00.

"This is an awakening book. Its plea is for designs in the surroundings of
houses. It insists that the house and the ground around the house should be
arranged in relation to each other. It maintains the irrefutable proposition that
really satisfying beauty in the immediate surroundings of men's lives upon this earth
must spring, not from any imitated likeness to wild nature, nor yet from any imprac-
ticable conformity to the ideals of landscape painters, but simply from the harmonious
adaptation of land and buildings to the uses and enjoyments of real life. . . . Our
private and public gardens, with their necessarily unnatural and yet studiously in-
formal arrangements, betray the same lack of feeling for design. Our time is certainly out of joint as respects this art, and for this reason this straightforward book is
peculiarly valuable and welcome." — *The Nation.*

BOTANY. — **Annals of Botany.**   Edited by J. B. BALFOUR, S. H. VINES,
and W. G. FARLOW.   Vols. I. and II., $5.75 each.   Volume I.
Half morocco, $9.00.   Subscription, per vol., $8.00.

BOWER (F. O.). — **A Course of Practical Instruction in Botany.**
> Second Edition.   Revised.   Complete in one volume. $2.60.
> Part I.   Phanerogamæ, — Pteridophyta. $1.50.

"Of the great value to both teacher and pupil of such a hand-book, there can
be but one opinion. . . . The method of paragraphing adopted permits of an
easy handling of the subject. The student is directed to look at certain parts and

is told some characteristic by which they may be recognized, while at the same time much information of a theoretical nature, points in homology, special methods of demonstrating a difficult feature, the suggestion of comparative studies, and various other helpful matters, are interspersed. Reagents and the latest processes of staining are freely used. The interpretation of structure according to the most recent investigations, and a corresponding nomenclature, are items that will be highly appreciated by the progressive student. The publication will enable English students to obtain a practical knowledge of the fundamental features of plant structure in accordance with the latest views, and it is, therefore, a much needed work." — *The Botanical Gazette.*

BRIGHT. — WORKS BY HENRY A. BRIGHT. **A Year in a Lancashire Garden.** $1.25.

"'A Year in a Lancashire Garden' possesses a peculiar charm for those fond of out-door gardening, and, in fact, for all those who love out-door life in any shape. It is a sort of monthly record of the experiences of the author among his plants and flowers, and details his plans and projects, his successes and failures. Much can be learned by those who make a specialty of plant raising from his pleasant pages. There is a pleasant literary flavor about the book, too, which will add to the enjoyment of the reader." — *The Boston Transcript.*

The English Flower Garden. $1.25.

CARPENTER. — **Vegetable Physiology, etc.** $1.75.

DE BARY. — WORKS BY DR. A. DE BARY. **Comparative Anatomy of the Vegetative Organs of the Phanerogams and Ferns.** , With Woodcuts. 8vo. $5.50.

"This work is of the utmost value to students of botany. It treats in a most exhaustive manner of the various forms of vegetable tissues, their origin, development, modifications, and various changes. It is a text-book on vegetable histology of the highest order, and it is safe to say there is no more valuable work of its kind extant." — *The Microscope.*

"The publication of this work in this form will do much to stimulate a more exact study of the minute anatomy and development of the various organs of plants, and will, we trust, do not a little toward placing botanical work in this country upon something of the same basis as that of zoölogy." — *The American Naturalist.*

The Morphology and Biology of Fungi, Mycetozoa, and Bacteria. With numerous Illustrations. 8vo. $5.50.

Lectures on Bacteria. Second improved Edition. With Wood Engravings. 8vo. $1.50.

EARLE (J.). — **English Plant Names from the Tenth to the Fifteenth Century.** 16mo. $1.00.

GAYE. — The Great World's Farm. Some Account of Nature's Crops and How they are Grown. By SELINA GAYE, author of "The World's Lumber Room." With a Preface by G. S. BOULGER, F.L.S., and numerous Illustrations. $1.50.

"One of the delightful semi-scientific books which every one enjoys reading and at once wishes to own. Such works present science in the most fascinating and enticing way, and from a cursory glance at paragraphs the reader is insensibly led on to chapters and thence to a thorough reading from cover to cover. . . . The work is especially well adapted for school purposes in connection with the study of elementary natural science, to which modern authorities are united in giving an early and important place in the school curriculum." — *The Journal of Education.*

GOEBEL (Dr. K.). — **Outlines of Classification and Special Morphology of Plants.** With 407 Woodcuts. 8vo. $5.25.

HANBURY (D.). — **Science Papers,** chiefly Pharmacological and Botanical. Edited by J. INCE, F.L.S. With Illustrations. 8vo. $4.50.

HAYWARD (W. R.). — **The Botanist's Pocket-book.** 6th Edition. Cloth, limp. $1.25.

HOLDEN (H. A.). — **Foliorum Silvula.** Part I. 12th Edition. $1.75. Part II. 4th Edition. $1.25.

Folia Silvulæ. Fasciculos III., IV. $1.10.

HOOKER (Sir J. D.). — **The Student's Flora of the British Islands.** 3d Edition, revised. 16mo. $3.00.

"This third edition gives evidence that the great popularity of this well-planned and well-executed Flora continues undiminished. It is not increased in size; the changes, though not inconsiderable in the treatment of certain groups and species, are not striking, but the revision has been evidently critical. *Subspecies* are more largely adopted: varieties accordingly play a diminished part, — a fair compromise between the schools of narrow and wide limitation of species, and perhaps a necessity in the long-worked Floras of Europe. 'Characters concerned in the process of fertilization' are also introduced into this edition." — *American Journal of Science.*

Index Kewensis: Plantarum Phanerogamarum Nomina et Synonyma omnium Generum et Specierum a Linnæo usque ad annum MDCCCLXXXV complectens. Sumptibus beati Caroli Roberti Darwin, ductu et consilio Josephi D. Hooker, confecit B. Daydon Jackson. Fasciculus I.

"Botanists may well be congratulated on the issue of the first part of this important work. It is not only destined to be in constant use by working botanists, but will also serve as a standard of nomenclature for a considerable time to come. . . .

"The present fasciculus of 728 pages brings us to the letters *Den.* These details will suffice to show the magnitude of the work. Its importance may further be exemplified by the circumstance that it is no mere compilation, such as could be effected by collating other lists and indexes, but an absolutely original work, each name being traced back to its initial source, and the axiom 'verify your references' most scrupulously acted on. Carried out as it has been, there is no doubt that the work will be, as Mr. Darwin desired it to be, 'of supreme importance to students of systematic and geographical botany and to horticulturists.'" — *The Academy.*

JOHNSON (G. W.). — **The Gardener's Dictionary.** New edition, revised. Large 12mo. In 8 parts. Each 40 cents.

LASLETT (T.). — **Timber and Timber Trees.** With Illustrations. $2.50.

LAURIE (A. P.). — **The Food of Plants.** An Introduction to Agricultural Chemistry. With Illustrations. 18mo. 35 cents.

LUBBOCK. — WORKS BY SIR JOHN LUBBOCK, F.R.S., D.C.L.

     On British Wild Flowers. With Illustrations. *Nature Series.* $1.25.

     Flowers, Fruits, and Leaves. With Illustrations. *Nature Series.* $1.25.

"This fascinating little book reads like a fairy tale for amusement, though it is really the simplest and most direct statement of the latest discoveries of scientific observation. One need not be a botanist, or care anything about botany, to read with pleasure this wonderful little tale of actual facts about flowers and leaves. It is simple enough to be clear to the merest child, and astonishing enough to send a thrill along the most blasé of nerves." — *The Critic.*

"Models of what popular scientific works should be, showing that exact and reliable scientific information can be imparted in a clear and interesting manner." — *Popular Science News.*

MASSEE (G.). — **The Plant World.** With 56 Illustrations. *Whittaker's Library of Popular Science.* 16mo. $1.00.

"A very attractively written book, well adapted to create a taste for the study of botany. It enlists the sympathy of the reader at once by treating the plant not as a dead thing, but as a living creature, whose characteristic manifestations of life are to be lovingly watched and tended." — *The Home Journal.*

"We are very much pleased with this choice little volume. The language is concise, classical, scientific, yet clear and easily understood. It is a beautiful introduction, as its author claims, to the study of botany. The comparison of plant and animal life, aided by powerful microscopes and apt illustrations and diagrams, makes the plants seem our relatives." — *The Western Garden.*

     British Fungus-Flora. A Classified Text-book of Mycology. In three volumes. Vols. I., II., and III. $2.00 each.

MÜLLER (H.). — **Fertilization of Flowers.** With Illustrations. 8vo. $5.00.

"A splendid illustration of the industry and patience with which eminent explorers in science pursue their investigations is afforded in Dr. Hermann Müller's voluminous work. It contains an enormous mass of observations concerning the mechanism of flowers, their adaptations, by means of structure, color, and odor, to secure cross or self-fertilization, and the species of insects which in each given case assist in the operation. The data are carefully systematized, and their bearing upon the theory of Darwin, that cross-fertilization is more productive of vigorous seed than self-fertilization, is clearly pointed out. No botanist has worked more faithfully than Dr. Müller along these lines of research, or has obtained richer results as the fruits of many studious years." — *The Dial.*

NICHOLLS (H. A. A.). — **A Text-book of** Tropical Agriculture. With Illustrations. $1.30.

NISBET (J.). — **British Forest Trees and** their Sylvicultural Characteristics. $2.50.

" Dr. John Nisbet served many years in the care of the teak forests of tropical Burma, and has brought to Britain all the wealth of his experience. . . The author confesses that it is, to a considerable extent, a compilation from the best German sources, but says that he can at any rate vouch for the correctness of the scientific principles enunciated, from his own observations in both Britain and Burma. . . . Of course the climatic conditions of the United States are as different from those of England as our kinds of forest trees are unlike those of Europe ; but the general principles of forestry set forth in this treatise may well be studied by thoughtful Americans." — *The Nation.*

OLIVER (DANIEL). — **Lessons in Elementary Botany.** With nearly 200 Illustrations. New Edition. 16mo. $1.10.

PETTIGREW (J. B.). — The Physiology of the Circulation in Plants, **in the Lower** Animals, and in Man. With Illustrations. 8vo. $3.00.

SACHS (Dr. JULIUS). — **Lectures on** the Physiology **of** Plants. With 455 Woodcuts. Royal 8vo. $8.00. — See also *Goebel.*

The History of Botany. 1530–1860. $2.50.

SMITH (J.). — **Dictionary of the Popular Names of Plants.** 8vo. $3.50.

". . . A very valuable botanical work by the ex-curator of the Royal Gardens at Kew, England. It includes all the plants made use of by man in any way, either for ornament or medicine, or any other purpose. The plants are arranged alphabetically under the popular name, if there is any, and if not, then under the Latin title. The definition gives the Latin name and a simple description of the plant, its habitat, and the uses to which it is put. Mr. Smith's high reputation guarantees the accuracy of the work." — *The Boston Advertiser.*

SMITH (W. G.). — **Diseases of Field and Garden Crops.** With Illustrations. 16mo. $1.50.

" In this compact little volume the author has brought together the notes of the course of lectures given at the request of the officers of the Institute of Agriculture at the British Museum, South Kensington. It was the endeavor of the author, we are informed, to keep three objects clearly in view, viz., '*first*, the description only of such diseases as are of economic importance; *second*, the definition of all the phenomena of the diseases, in familiar words, such as, with proper attention, may be understood by all; this has been done without sacrificing scientific accuracy, as all botanical terms in common use are adverted to and explained : *third*, the consideration of the best means of preventing the attacks of plant diseases.' An examination of the book warrants us in saying that the author has succeeded admirably in his attempt. Every progressive farmer and gardener ought to procure this book and give it a careful reading." — *The American Naturalist.*

STRASBURGER (E.). — **Handbook of Practical Botany.** Edited from the German by W. HILLHOUSE. 3d Edition. With numerous Illustrations. 8vo. $2.50.

SOLMS-LAUBACH. — An Introduction to **Palæophytology from the** Standpoint of the Botanist. 8vo. $4.50.

SOWERBY'S English Botany, containing a Description and Life-size Drawing of every British Plant. Edited by T. BOSWELL. 3d Edition, entirely revised.

Vol.				Vol.			
I.	7 Parts,	161 Plates,	$12.00	VIII.	10 Parts,	208 Plates,	$18.00
II.	7 "	165 "	12.00	IX.	7 "	161 "	12.00
III.	8 "	169 "	16.00	X.	7 "	140 "	12.00
IV.	9 "	193 "	18.00	XI.	6 "	139 "	12.00
V.	8 "	183 "	16.00	XII. With an Index to the whole Work,			
VI.	7 "	160 "	12.00	**6 Parts,**	98 Plates,	$11.00	
VII.	7 "	160 "	**12.00**				

TANNER. — WORKS BY HENRY TANNER, F.C.S.

First Principles of Agriculture. 18mo. 30 cents.

". . . One of the books which should be in the hands of every farmer who wishes to understand the 'whys and wherefores' of his business. The author is recognized as one of the ablest writers on agricultural science in England, and he has contrived to pack into the ninety-five pages of this little primer a large amount of matter of the utmost value to the practical farmer, expressed in so simple language that a child may understand it. As far as possible, the use of technical terms has been avoided, and the few that are employed are carefully explained. The subjects treated are such as the farmer constantly has to deal with, but which too generally he deals with in a blind and groping way, when by the careful study of a simple text-book like this he would be able to act intelligently, and hence far more successfully, than without such aid." — *The Examiner.*

The Abbot's Farm ; or, Practice with Science. 90 cents.

"The author has woven the latest discovered facts and principles of scientific agriculture into a very readable story of a farmer's life as it might be made to be at the present time under English laws, and at ordinary rates for rent. . . . Agricultural subjects now most engrossing the attention of thinking farmers are all discussed candidly, and at the same time the interest of the story goes along with the subjects." — *The Country Gentleman.*

" . . . An attractive volume that every farmer should read and consider. It is a good answer to the question, Does instruction in science make a man a more successful farmer and a better neighbor ? and one might look long and not find so practical and satisfactory an answer as these charming and pleasantly written sketches furnish the reader." — *The Boston Traveller.*

The Alphabet of the Principles of Agriculture. Being a First Lesson-Book on Agriculture. 16mo. 15 cents.

Further Steps in the Principles of Agriculture. 30 cents.

Elementary School Readings in the Principles of Agriculture for the Third Stage. 30 cents.

Elementary Lessons in the Science of Agricultural Practice. 16mo. 90 cents.

VINES (S. H.). — **Lectures on the Physiology of Plants.** 8vo. $5.00.

" Dr. Vines . . . in this handsome and handy volume has now given the general scientific or scientifically minded reader easy access to the principles of vegetable physiology in their present aspect. The arrangement and the presentation are admirable, the style clear and good, and the book, unlike some translations from the German into English, may be read with satisfaction. It may be commended even to the higher order of our botanical teachers, not only as an excellent model, but also for fulness of information and clear indication of the actual lines of research. Each lecture, moreover, is supplemented by a select bibliography of the topic." — *The Nation.*

WARD (H. M.). — **Timber and Some of its Diseases.** With Illustrations. *Nature Series.* $1.75.

" A praiseworthy work in every respect. . . . It is by all odds the clearest *précis* in the field. . . . The book will be of use in the hands of every one who is interested in the care of trees, and especially to the students in our agricultural schools, to whom must be largely intrusted the intelligent supervision of our forests." — *The Evening Post.*

WRIGHT (J.). — **Horticulture.** Ten Lectures delivered for the Surrey County Council. 18mo. 35 cents.

" This ' Primer of Horticulture ' is designed as an introduction to a scientific and practical study of gardening and fruit growing, either for the small householder, who enjoys the care of his seven-by-nine piece of ground, or for the farmer to whom the best and most economical methods are matters of ' dollars and cents.' . . . The construction of the work is admirable, and it might be read with profit by many scientific men as a model for popular scientific exposition. Great care has been taken to select the most important aspects of the topic discussed, the essential facts being presented in clear and untechnical language, while the subject is not overburdened with detail." — *Popular Science Monthly.*

MACMILLAN & CO.,
66 FIFTH AVENUE, NEW YORK.

www.ingramcontent.com/pod-product-compliance
Lightning Source LLC
Chambersburg PA
CBHW030556270326
41927CB00007B/939